D1413104

- **ENERGY**
  - **SIMULATION-TRAINING**
    - **OCEAN ENGINEERING AND INSTRUMENTATION**

Research Papers of the Link Foundation Fellows

Volume 1

Brian J. Thompson
Editor

Published by
The University of Rochester Press
in Association with
The Link Foundation

First published 2001

University of Rochester Press
668 Mt. Hope Avenue
Rochester, New York, 14620, USA

and at

PO Box 9, Woodbridge,
Suffolk IP12 3DF, UK

ISBN 1–58046–093–3

Library of Congress Cataloging-in-Publication Data

Energy, simulation-training, ocean engineering and instrumentation : research papers of the Link Foundations Fellows / Brian J. Thompson, editor.
p. cm.
ISBN 1-58046-093-3 (alk. paper)
1. Power resources   2. Energy development.   I. Thompson, Brian J., 1932–

TJ 163.2.E4867 2001
333.79—dc21   2001021331

This publication is printed on acid-free paper
Designed and Typeset by Isis-1 Corporation
Printed in the United States of America

The editor and the authors dedicate this volume to the memory of Edwin A. Link and his innovative and entrepreneurial spirit.

TABLE OF CONTENTS

PREFACE

An important ingredient of education at the Ph.D. level is the reporting of the original scholarly and intellectual content of the research conducted by the candidate for the degree. Thus the Link Foundation asks that each recipient of a Link Foundation Fellowship prepare and submit a journal length article resulting from their work under the Link Foundation sponsorship. The Link Foundation believes in this discipline and has decided to complete this process by the formal publication of these papers in a single volume.

The current fellowship programs are in the fields of energy, simulation-training and ocean engineering and instrumentation. The Energy fellowships have been offered since 1983 and the papers of these fellows have been published in a series of fifteen volumes. Simulation-Training fellows have submitted research papers since the inception of the program in 1990; these papers have been published informally. Finally the program in Ocean Engineering and Instrumentation was launched in 1998 and the first fellowship papers are published in the volume.

The Link Foundation is pleased to have been able to support the students represented here. The work reported covers a wide variety of research topics carried out at leading universities and colleges.

Brian J. Thompson
University of Rochester and
The Link Foundation
Editor

PART I

ENERGY

Nitric Oxide Measurements in Jet Flames with a Ti:Sapphire Laser

Michael W. Renfro

School of Mechanical Engineering

Purdue University

West Lafayette, IN 47907–1288

Research Advisors: Dr. Normand M. Laurendeau and Dr. Galen B. King

ABSTRACT

Measurements of species concentrations in flames are important to understanding the detailed structure of the reactions that produce pollutants such as nitric oxide (NO). Many groups have demonstrated direct measurements of NO by using laser-induced fluorescence (LIF), usually with pumped dye lasers. In this work, a Ti:Sapphire laser is used for the first time (to our knowledge) to measure NO concentrations in flames. This procedure involves frequency quadrupling the laser output to access the necessary UV wavelengths. The motivation for this work is an attempt at time-series measurements of NO so that the frequencies of pollutant fluctuations in turbulent flames can be resolved. Time-series measurements of other species, such as OH and CH, have been previously demonstrated with the same laser system. Measurements of time-averaged NO and time-series measurements in unsteady laminar flames are presented using the new system. Measurements of fluorescence lifetimes for NO are also recovered and compare favorably to predictions using standard quenching cross-sections. However, the Ti:Sapphire system does not presently provide enough power for full time-series measurements of NO at elevated Reynolds numbers. The technique for using a Ti:Sapphire laser for NO studies and necessary modifications for extension to fully turbulent flames are discussed.

INTRODUCTION

Combustion provides over 80% of all energy consumed in the U.S. [1]. While nuclear and hydroelectric power have become more important in electric plants, coal combustion still provides over 50% of the United States' electricity, and direct use of combustion in industrial manufacturing, domestic heating, and transportation applications is abundant. Thus, even small gains in conversion efficiency or small reductions in pollutant production per unit fuel can have an enormous economic and environmental impact. However, combustors are often designed to provide large energy outputs from a small physical device. Moreover, rapid mixing of fuel and oxidizer is usually desired. As a result of these design goals, most practical combustors utilize turbulent flow [1]. In this context, a quantitative understanding of turbulent combustion is a clearly motivated goal.

In any turbulent flow, the velocity and scalars fluctuate in a manner such that their values cannot be specified with certainty [2]. Traditionally, it is only the mean quantities that are of engineering interest, such as the total heat production rate or total pollutant output. Hence, many models of reacting turbulence only predict time-averaged quantities. Unfortunately, these fluctuations can give rise to complicated turbulence-chemistry interactions because of a strong nonlinear dependence of the chemical production rates on temperature. In this case, the time-averaged production rate of some species (say the pollutant nitric oxide) cannot itself be computed from other time-averaged quantities; details of the fluctuations must be understood and must be included in the turbulent combustion model. Advanced models are being developed based on statistical approaches, which can recover significant information about the turbulent fluctuations, but it is imperative that these models be tested against measurements in benchmark flames. These measurements must eventually recover multiple species concentrations, multiple spatial points, and multiple temporal points. Hence, unique measurement that go beyond time-averaged data are required to improve turbulent combustion models thereby providing better tools for the design of cleaner, more efficient, energy conversion devices.

Several diagnostic techniques meeting this goal have already been developed. For example, single-shot planar laser-induced fluorescence (PLIF) techniques [3–5] can measure the instantaneous, two-dimensional structure of a single species within a turbulent flame. However, PLIF does not typically recover multi-time information, and is usually limited to one or a few scalars. Likewise, the single-shot, simultaneous measurements of hydroxyl (OH), nitric oxide (NO), temperature, and major-species concentrations by Barlow and coworkers [6,7] recover multi-scalar data at a single-point but do not yield spatial or temporal information.

In short, measurements of multiple species concentrations at a single point and a single time are possible, as are measurements of a single species at multiple points and a single time. To complement these techniques, a laser diagnostic technique termed picosecond time-resolved laser-induced fluorescence (PITLIF) was developed which is capable of quantitative time-series measurements of minor-species concentrations [8–11]. This measurement provides a single-species concentration at a single point but multiple times. Since this technique does not capture multiple spatial points or multiple scalars, it is also limited; nevertheless, the temporal information of minor species included has been previously unexamined and provides a new outlook on the structure of fluctuations in turbulent flames. Techniques for the measurement of time series for temperature [12] and velocity [13] are well established and studies of their multi-time statistics have been published. The extension of time-series measurements to minor-species concentrations, which are important to understanding detailed flame structure and especially pollutant formation, was not possible prior to PITLIF.

PITLIF has been recently demonstrated for the quantitative measurement of OH and methylidyne (CH) time series in turbulent $H_2/CH_4/N_2$ nonpremixed jet flames [14,15]. This flame has also been used to study temperature time series [16] and is presently being studied for mixture fraction and radiation statistics. Similar flames have been studied at both the Technische Universität Darmstadt, Germany [17] and at Sandia National Laboratories [18]. Multi scalar, single-point data are available for this fuel mixture and geometry via the internet, as are single scalar, multi-point PLIF images [17–19]. Thus, this flame has been studied by each of the complimentary techniques (one for multiple species, one for multiple points, and one for multiple times), making it perhaps the most thoroughly characterized turbulent nonpremixed flame in the literature.

This report details the first attempt to extend the PITLIF technique to a direct measurement of NO, an atmospheric pollutant. This measurement complements the previous CH and OH work since the three species display markedly different chemical time scales and should be susceptible to the effects of turbulent-chemistry interactions to varying degrees. Some differences between CH and OH time-series statistics have been previously found, and it is hoped that similar measurements of NO will amplify these differences and lead to a better understanding of the effects of turbulence-chemistry interactions. Presently, limitations in the signal-to-noise ratio (SNR) prohibit recovery of clean autocorrelation functions and integral time scales, the two statistics for which differences were noted between OH and CH measurements. Nevertheless, time-averaged concentrations are reported from a series of hydrogen-based jet flames and time-series measurements in periodic laminar flames are demonstrated. To our knowledge, these are the first measurements of NO using a Ti:Sapphire laser. As these lasers are now widespread, elements of the technique described here may be useful for permitting studies of NO by laboratories not previously equipped to do so. In addition, potential improvements to the technique that would eventually allow recovery of time-scale statistics are discussed.

METHOD

PITLIF Laser System for NO Excitation

PITLIF is a linear laser induced fluorescence technique. As such, a laser is tuned to excite a molecule in some prechosen rotational, vibrational, and electronic (rovibronic) configuration to a higher-energy rovibronic configuration. The laser wavelength corresponds to the difference in the two energy levels. Once excited, the molecule will decay back to a lower rovibronic level via either fluorescence (~2% probability for our case) or electronic quenching (~98% probability). For the fluorescence events, photons are released in all directions, and the rate of these photon emissions is linearly related to the concentration of molecules in the ground state, the laser power, and the fluorescence lifetime (a measure of the rate of relaxation by quenching). A diagram of the PITLIF laser system, including the burner station, is shown in Fig. 1. Full details of the PITLIF system are available elsewhere [8,10,20]. Briefly, a continuous argon-ion multi-mode laser (20 W) pumps a Spectra Physics Tsunami, regeneratively mode-locked, Ti:Sapphire laser. The Tsunami output is an 80–MHz repetition rate train of infra-red (IR) pulses, which have a continuously-tunable wavelength from 840 to 1000 nm. The pulse width can be switched between 1.5 and 18 ps by changing an optic in the Tsunami cavity. After leaving the Tsunami, the IR beam was frequency doubled to 452 nm in a CSK SuperTripler. The conversion process is non-linear and proportional to the peak power squared [21]. For the previous OH and CH measurements [14,15], the output of the SuperTripler was used directly. For NO studies, a wavelength of ~226 nm (see next section) is required and this cannot be achieved by just doubling or tripling the Tsunami output. Thus, the 452–nm beam was redoubled in an external BBO crystal (Quantum Technology BBO, 5x5x6 mm^3, cut at 60°) to achieve the required wavelength. This crystal is placed at the focus of an external lens to increase the beams intensity and hence the doubling efficiency. The laser beam, now at a frequency tuned to excite NO, was recollimated by another lens and focused by a 22.9–cm focal length, 5.1–cm diameter UV lens through the probe volume, which lies above the burner. The laser beam was then dumped into a photodiode, which recorded the average laser power.

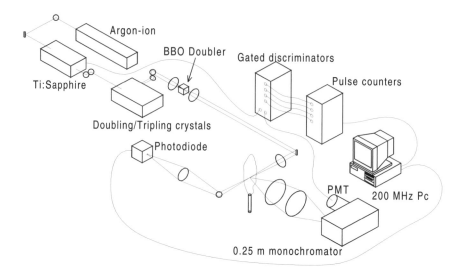

Figure 1. Experimental setup for the PITLIF system. KrF mirrors (CVI Model No. KRF-1025–0) centered at 226 nm were used for the NO studies, and a second doubling crystal was inserted after the CSK SuperTripler, as compared to the previous CH measurements [15].

Spectroscopic Line Selection for NO Excitation

In order to relate the fluorescence signal (number of collected fluorescence photons) to the item of interest, in this case the total concentration of NO, it is necessary to pick a ground rovibronic state that has a uniform population and absorption cross-section as a function of temperature. A simulated absorption spectrum is shown in Figure 2 for NO excitation with the 2–ps pulse. This simulation uses the code developed by Seitzman [22]. As for CH, the rovibronic lines for NO are very close in the (0,0) vibrational band of the X $^2\Pi$—A $^2\Sigma^+$ electronic system. Since NO requires frequency quadrupling of the Tsunami IR output, the usable UV laser power (and hence the signal) depends even more on IR power than for the tripling process used for CH. Thus, operation of the Tsunami with a 2–ps pulse was chosen, as this increases the power at the peak of the laser pulse. From a separate spectral simulation with 18–ps excitation, the peak absorption was found to be only three times larger than that for the 2–ps pulse. However, the laser energy is expected to be at least an order-of-magnitude larger with the 2–ps pulse. Thus, the chosen excitation location, shown in Figure 2 at 225.06 nm, gives the best signal and temperature independence (8% from 1000 to 2500 K) available with the Tsunami laser. For 2–ps excitation, 10 P-, Q-, and R-branch rotational lines are excited in the range 225.02–225.09 nm.

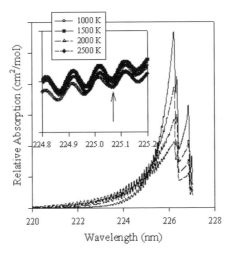

Figure 2. Simulated NO absorption spectrum. The inset details the region for which the temperature dependence is minimized. The arrow denotes the chosen excitation location.

Standard PITLIF Detection System with Gated Photon Counting

For previous OH and CH measurements, the fluorescence signal is collected at 90° from the incident laser beam and focused through a 0.25–m monochromator and onto a Hamamatsu HS5321 fast-response photomultiplier tube (PMT) as shown in Figure 1. The PMT converts each fluorescence photon into an analog electric pulse (~20–mW peak, 1–ns duration). These pulses are converted to NIM logic pulses using a leading edge discriminator, and are split into four identical signals. These four signals are separately delayed and gated to determine the photon count rate for separate temporal delays relative to the laser beam [10]. The gated signals were counted by four parallel pulse counters. This gated photon counting system was utilized to measure, simultaneously, the fluorescence signal and the fluorescence lifetime (which is a decay constant of the order 2 ns). By considering both values, a quantitative measure of the species concentration can be recovered [23,8]. Similar measurements of NO fluorescence using this photon counting system were attempted and are reported here. However, measurements of nitric oxide are not possible in turbulent flames by using the standard PITLIF procedure because of signal-to-noise ratio limitations. This limitation arises because of the need to frequency quadruple the Tsunami laser beam to generate NO excitation wavelengths. The maximum power achievable is approximately 5 mW, which is a factor of 100 less than that for CH. Moreover, NO concentrations are only about a factor of 10 greater than those for CH in the primary $H_2/CH_4/N_2$ flames

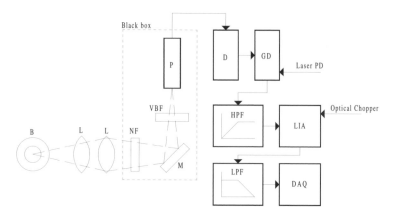

Figure 3. Experimental setup utilized for measurements of time-averaged NO fluorescence. B: burner station; L: fused-silica lens; NF: 226–nm notch filter; M: Excimer laser cavity mirror; VBF: visible-light blocking filter; P: photomultiplier tube with attached slit; D: discriminator; GD: gated discriminator; PD: photodiode; HPF: high-pass analog filter; LIA: lock-in amplifier; LPF: low-pass analog filter; DAQ: data acquisition computer.

investigated here. This reduction in fluorescence signal yields a signal-to-background ratio (SBR) of less than $1/10$ and prohibits clean measurements of NO time series. Thus, an alternate instrumentation setup was designed to permit time-averaged measurements of NO fluorescence profiles in turbulent flames and time-series measurements in unsteady laminar flames.

Alternate Detection Instrumentation for NO Measurements

Figure 3 shows the alternate instrumentation used for the NO measurements. The PMT was removed from the 0.25–m monochromator used for the CH and OH measurements and was placed after the two fluorescence collection lenses. A notch filter was used to remove Rayleigh or particle scattering at the laser wavelength [24]. The fluorescence was directed to the PMT using a bandpass mirror which reflects the (0,0) through (0,6) vibrational bands of NO but passes visible light. A slit before the PMT defines the probe volume dimensions and was fixed for a 1–mm probe length for all NO measurements. These changes from the OH and CH setup permit a higher fluorescence signal by increasing the throughput of the optical system. However, it was necessary to add an additional spectral filter before the PMT, as shown in Figure 3, to block the residual flame emission reflected by the bandpass mirror and reflected off nearby objects. The total signal collected with this setup is close to that of the monochromator system but with a better SNR.

The same photon-counting system was used to convert the anode pulses from the PMT into logic pulses. The amplitude of the pulses from the PMT are approximately 20 mV (when 50 W terminated) and fluctuate from pulse to pulse because of shot noise at the dynode stages of the PMT. In other words, the pulse train from the PMT is weak and noisy. Following discrimination, each pulse has the same height, which eliminated dynode noise, and is amplified by a factor of about 50. The output from this discriminator was filtered by a high-pass filter set at 6000 Hz and amplified using a modified lock-in amplifier. The lock-in amplifier was referenced to a 10–kHz optical chopper placed in the laser path before the second, external doubling crystal used for NO excitation.

The output from the lock-in amplifier was monitored following its mixer but before its low-pass filters. This procedure is necessary since the built-in amplifier filters are too slow for potential time-series measurements, as discussed by Renfro *et al.* [11] who used the same lock-in amplifier for CH measurements. An external low-pass filter followed the lock-in amplifier. The same data acquisition hardware and software used by Renfro [25] was employed to sample the resulting signal. For some of the measurements discussed in this report, the photon counting system was used for comparison to the lock-in amplifier measurements.

RESULTS

Measurements of Cold NO in Nitrogen/Oxygen Jets

Unlike many other species of interest for flame measurements, NO is stable at room temperature so that initial studies were undertaken in a simple nonreacting jet of bottled NO mixed with nitrogen at room temperature. The NO/N_2 mixture was diluted with either more nitrogen or oxygen to change both the NO concentration and the NO fluorescence lifetime. Nitrogen is a poor quencher for NO, thus a mixture of NO in only nitrogen has a long fluorescence lifetime and a proportionally large total fluorescence signal. A detection scan of NO fluorescence for 300 ppm of NO in N_2 is shown in Figure 4 using the standard PITLIF system (Figure 1). For this measurement the excitation wavelength was fixed as discussed in the previous section. The monochromator wavelength was then scanned. The resulting plot shows the distribution of the fluorescence signal among the different vibrational transitions permitted from the excited state. The same measurement is shown with pure nitrogen at the probe volume to compare the NO fluorescence signal to that of Rayleigh scattering at the laser wavelength. This scattering signal can be much larger than the NO fluorescence signal under flame conditions where the fluorescence lifetime is short. For this reason, the narrow-band mirror described in Figure 2 was used to reject Rayleigh scattering for the alternate detection system.

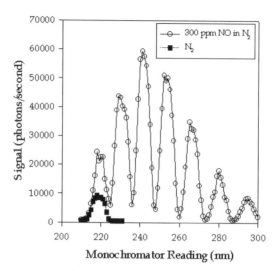

Figure 4. Detection scan of NO fluorescence at room temperature. The laser wavelength is fixed and the monochromator wavelength is altered. The first six vibrational band transitions are clearly apparent. A separate detection scan without NO is shown to compare the signal strength of Rayleigh scattering.

With the PITLIF system, the fluorescence lifetime of NO was measured as a function of the oxygen content of the bath gas. Based on the quenching correlations of Paul et al. [26], the fluorescence lifetime can be predicted for the same conditions at a temperature of 300 K. A comparison of this prediction and the measured lifetime is shown in Figure 5. The comparison is quantitatively very good, although the data points display some scatter.

Based on these results, it is clear that the Ti:Sapphire laser system can be utilized for measurements of NO and fluorescence lifetimes can be recovered. However, the effects of lower signals (by a factor of 50) owing to the reduced fluorescence lifetime in a flame and the increased noise background from flame emission makes the SNR very poor for flame measurements. For this reason, the alternate detection system was used to improve the SNR for the remaining results presented here. However, this system was not used to measure fluorescence lifetimes, thus these remaining results are qualitative measures of the NO concentration itself.

Time-Averaged NO Measurements in Jet Flames

Radial profiles of NO fluorescence are shown in Figure 6 for a jet flame with a fuel composition of 22.1% CH_4 + 33.2% H_2 + 44.7% N_2 by volume (designated flame A3), flame A3 without any nitrogen, and flame A3 without methane or nitrogen (pure hydrogen) at various axial heights and

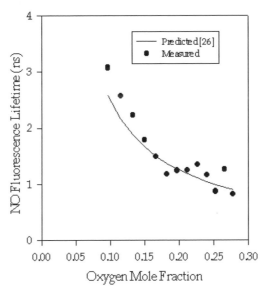

Figure 5. Measured and predicted [26] fluorescence lifetime for NO issuing from a jet in a 300 K bath of nitrogen. Measurements are shown for varying levels of dilution of the NO/N_2 mixture with O_2.

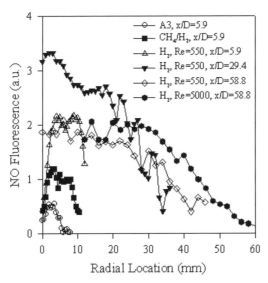

Figure 6. Radial profiles of NO fluorescence in various hydrogen-based nonpremixed jet flames. Each flame issues from a burner with a 3.4–mm diameter. Flame A3 is a mixture of 22.1% CH_4, 33.2% H_2, and 44.7% N_2 as used in previous OH and CH studies [14,15]. The CH_4/H_2 flame has the same fuel composition as A3 without nitrogen.

Reynolds numbers. Each point represents the time-average of 50 seconds of data with a low-pass filtering frequency of 1 Hz (the standard lock-in amplifier filters were used in this case). The increase in NO is substantial as the nitrogen diluent is removed and again as the methane is removed. In both cases this likely occurs because the flame temperature has increased, permitting the production of more thermal NO. In all three flames, the profile remains bimodal with a local minimum in NO concentration on the jet centerline.

Radial profiles of time-averaged NO fluorescence are included in Figure 6 for a laminar Re=550 hydrogen jet at axial heights of 100 mm and 200 mm. The NO profile becomes unimodal at x=100 mm and is significantly wider than at the lower axial locations. At x=100 mm, the peak NO concentration is much greater than that lower in the flame. This likely occurs because of the increased residence time for NO production. At x=200 mm, the NO profile is still wide, but the peak concentration has decreased by almost 50%. This is likely due to dilution of the flame by entrained room air.

Time-Series Measurements of Nitric Oxide

Measurement of NO time series were made using the alternate detection system of Figure 3. A Power Spectral Density (PSD) obtained from such a time series at the centerline for the highest temperature flame (pure hydrogen, Re=550) at x=100 mm and x=200 mm is shown in Figure 7. The time-series measurements were obtained with a low-pass filter set at 3 kHz. At the low Reynolds number of this jet, a large single-frequency oscillation is observed in the PSD corresponding to buoyancy-induced flicker in the laminar flame. Other than this single frequency, the PSD is flat at all frequencies for an axial height of x=100 mm. However, at x=200 mm, the second harmonic of the buoyancy-induced oscillation is clearly visible. This behavior has been observed for other minor-species concentration measurements in laminar jets [9,11]. The larger height also displays fluctuations over a broad range of scales at lower frequencies, with a clear decrease in fluctuation intensity for increasing frequency. This behavior was also observed for OH in laminar jet flames [25]. At both axial heights in Figure 7, the PSD is invariant with radial location as similar statistics are observed for the jet centerline and for larger radial locations.

Figure 8 shows the progression of PSDs measured at the jet centerline for x=200 mm as a function of Reynolds number. At Re=550, a broad PSD is apparent as in Figure 7, but the intensity of the broad-band fluctuations decreases with rising Reynolds number until, at Re=1030, only a small single-frequency component is apparent. For fully turbulent jets (Re>5000), the PSD was found to be completely flat for all measurement locations and conditions. This result occurs because the shot noise from NO fluorescence

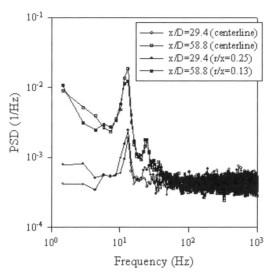

Figure 7. Power spectral densities of NO fluorescence in a laminar hydrogen diffusion flame (Re=550). The strong oscillation at ~10 Hz arises from buoyancy-induced oscillations and has previously bee observed with both OH and CH time-series measurements [14,15].

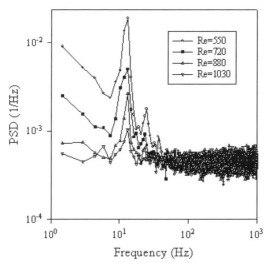

Figure 8. PSDs of NO fluorescence as a function of Reynolds number at x=200 mm on the centerline of several hydrogen jet diffusion flames. Each flame issues from a burner with a 3.4–mm diameter.

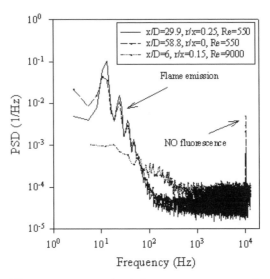

Figure 9. PSDs of flame emission and NO fluorescence measured with the original PITLIF photon-counting instrumentation. The NO fluorescence is shifted to 10.06 kHz by modulating the laser beam with an optical chopper.

and background flame emission is larger than any real NO fluctuations [25]. Thus, the flat noise spectrum dominates the measured spectra.

Another view of this noise progression is apparent when using the original photon-counting equipment, but with the 10–kHz chopper left in the laser path. Figure 9 shows the PSDs measured in this manner for two laminar and one turbulent jet. The spectrum for flame emission is apparent up to ~100 Hz for the laminar jets and 1000 Hz for the turbulent jet. For the flickering laminar flames, several harmonics of the buoyancy-induced oscillation are apparent. Because of the chopper in the laser path, the NO fluorescence signal is shifted to 10.06 kHz and is clearly visible in the photon-counting spectra.

Figure 10 shows the same spectra as in Figure 9 but only for the frequency region near 10 kHz. For the two laminar jets, the DC component of the NO signal is apparent as a large intensity at 10.06 kHz. For the downstream measurement at Re=550, the ~10 Hz buoyancy-induced oscillation appears as two sidebands to the DC component. The second harmonic is also visible at this measurement location. Comparing the x=200–mm and x=100–mm locations, the broadband low-frequency fluctuations discussed for Figure 7 are apparent as a broadened spectrum at 10 kHz (x/D=58.8).

Unfortunately, in the turbulent jet, even the DC component of the NO signal at 10 kHz is obscured by the shot-noise floor from the large flame emission signal. Essentially, the lock-in amplifier recovers the PSD of NO

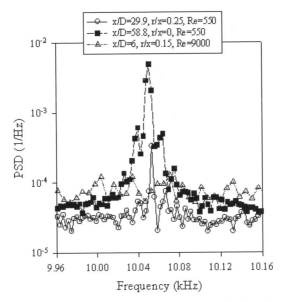

Figure 10. PSDs of NO fluorescence measured with the photon-counting instrumentation. The DC component of the NO fluctuations appears at 10.06 kHz because of modulation by the optical chopper. Sidebands from the primary peak are a result of the ~10 Hz oscillations apparent in Fig. 7. The lack of any spectral content for the turbulent flame is a result of shot noise in the weak signal.

concentration that is shifted to 10 kHz by modulation. Since even the DC component is not apparent at 10 kHz, the PSD for NO cannot possibly be obtained for a turbulent flame under these SBR conditions. With sufficient filtering (averaging), it is still possible to obtain time-averaged data as evidenced by the decent radial profile for a Re=5000 jet in Figure 4. For turbulent, pure-hydrogen jets, the SBR for NO is never better than 1:10 (compared to a maximum 1:2 for the Re=550 jet at x=200 mm). As observed in Figure 6, the NO signal is even smaller for the primary $H_2/CH_4/N_2$ flames, such that no reasonable modifications to the PITLIF system could recover spectra under these conditions.

SUMMARY

The measurement of NO fluorescence in a flame has been demonstrated for the first time (to our knowledge) using a Ti:Sapphire laser system. This could be important for future studies of NO as these laser systems are becoming common in laboratories for many fields of study. The modifications required to access NO wavelengths include frequency quadrupling the laser beam using two doubling crystals. In our case, a relatively low powered laser system with a high 80–MHz repetition rate was used. Hence, the maximum achievable laser power after frequency-quadrupling was about 5 mW at 226 nm (higher powers could be achieved but with significant damage to the BBO crystal). This power cannot be improved with the current instrumentation, and as apparent from the PSDs in the previous section, the high-repetition rate system is not sufficient for temporal scale measurements of NO in flames with fully turbulent Reynolds numbers. It may be possible to improve NO measurements through the use of an amplified Ti:Sapphire laser system which can provide pulse energies on the order of 1 mJ at UV wavelengths with laser repetition rates still in excess of 50 kHz. However, this alternative is neither trivial nor inexpensive, and may still yield marginal results for low temperature nitrogen-diluted flames. Nevertheless, for time-averaged measurements, particularly in laminar flames, the Ti:Sapphire laser system can be used for studies of NO concentrations.

Another possibility for time-scale recovery exists in the measurement technique of Yaney et al. [27] for which the autocorrelation function is directly measured from two separate laser pulses (delayed relative to each other in a controlled manner). The laser system in this case does not need to have a high-repetition rate so that two 10–Hz Nd:YAG-pumped dye laser systems are sufficient for time-scale recovery. While a complete autocorrelation function can be recovered, this technique will not recover a full time series, which contains phase information beyond the PSD or autocorrelation function.

ACKNOWLEDGEMENTS

I would like to acknowledge my co-major advisors Professors Normand M. Laurendeau and Galen B. King for their direction in the course of my Ph.D. research. Professor Jay P. Gore has also contributed significantly to the project. The support of the Link Foundation Energy Fellowship for this work is gratefully acknowledged. The Air Force Office of Scientific Research and the Department of Defense NDSEG Fellowship Program provided financial assistance for earlier parts of this research program.

REFERENCES

[1] S. R. Turns, "An Introduction to Combustion—Concepts and Applications," McGraw-Hill, Inc., New York (1996).
[2] H. Tennekes and J. L. Lumley, "A First Course in Turbulence," The MIT Press, Cambridge, MA (1972).

[3]   M. G. Allen, R. D. Howe, and R. K. Hanson, "Digital imaging of reaction zones in hydrocarbon-air flames using planar laser-induced fluorescence of CH and C_2," *Opt. Let.* 11, 126–128 (1986).

[4]   N. T. Clemens, P. H. Paul, and M. G. Mungal, "The structure of OH fields in high Reynolds number turbulent jet diffusion flames," *Combust. Sci. and Technol.* 129, 165–184 (1997).

[5]   M. C. Drake and R. W. Pitz, "Comparison of turbulent diffusion flame measurements of OH by planar fluorescence and saturated fluorescence," *Exp. in Fluids* 3, 283–292 (1985).

[6]   R. S. Barlow, R. W. Dibble, J.-Y. Chen, and R. P. Lucht, "Effect of Damköhler number on superequilibrium OH concentration in turbulent nonpremixed jet flames," *Combust. Flame* 82, 235–251 (1990).

[7]   R. S. Barlow and C. D. Carter, "Raman/Rayleigh/LIF measurements of nitric oxide formation in turbulent hydrogen jet flames," *Combust. Flame* 97, 261–280 (1994).

[8]   M. W. Renfro, G. B. King, and N. M. Laurendeau, "Quantitative hydroxyl-concentration time-series measurements in turbulent nonpremixed flames," *Appl. Opt.* 38, 4596–4608 (1999).

[9]   M. W. Renfro, S. D. Pack, G. B. King, and N. M. Laurendeau, "Hydroxyl time-series measurements in laminar and moderately turbulent methane/air diffusion flames," *Combust. Flame* 115, 443–455 (1998).

[10] S. D. Pack, M. W. Renfro, G. B. King, and N. M. Laurendeau, "Photon-counting technique for rapid fluorescence decay measurements," *Opt. Lett.* 23, 1215–1217 (1998).

[11] M. W. Renfro, M. S. Klassen, G. B. King, and N. M. Laurendeau, "Time-series measurements of CH concentration in turbulent CH_4/air flames by use of picosecond time-resolved laser-induced fluorescence," *Opt. Lett.* 22, 175–177 (1997).

[12] R. W. Dibble and R. E. Hollenbach, "Laser Rayleigh thermometry in turbulent flames," *Proc. Combust. Inst.* 18, 1489–1499 (1981).

[13] I. Gökalp, I. G. Shepherd, and R. K. Cheng, "Spectral behavior of velocity fluctuations in premixed turbulent flames," *Combust. Flame* 71, 313–323 (1988).

[14] M. W. Renfro, W. A. Guttenfelder, G. B. King, and N. M. Laurendeau, "Scalar time-series measurements in turbulent CH_4/H_2/N_2 nonpremixed flames: OH," *Combust. Flame*, in press (2000).

[15] M. W. Renfro, G. B. King, and N. M. Laurendeau, "Scalar time-series measurements in turbulent CH_4/H_2/N_2 nonpremixed flames: CH," *Combust. Flame* 122, 139–150 (2000).

[16] A. Lakshmanarao, "Time-series measurements of mixture fraction and temperature in turbulent non-reacting jets and turbulent nonpremixed flames," M.S. Thesis, Purdue University, West Lafayette, IN (1999).

[17] V. Bergmann, W. Meier, D. Wolff, and W. Stricker, "Application of spontaneous Raman and Rayleigh scattering and 2D LIF for the characterization of a turbulent CH_4/H_2/N_2 jet diffusion flame," *Appl. Phys. B* 66, 489–502 (1998).

[18] W. Meier, R. S. Barlow, Y.-L. Chen, and J.-Y. Chen, "Raman/Rayleigh/LIF measurements in a turbulent CH_4/H_2/N_2 jet flame: experimental techniques and turbulence-chemistry interaction," *Combust. Flame*, in review (2000).

[19] W. Meier, Personal communication (2000).

[20] M. W. Renfro, "Quantitative time series for minor-species concentrations: measurements and modeling in turbulent nonpremixed flames," Ph.D. Thesis, Purdue University, West Lafayette, IN (2000).

[21] S. A. Akhmanov and R. V. Khokhlov, "*Problems of Nonlinear Optics (Electromagnetic Waves in Nonlinear Dispersive Media),*" Gordon and Breach Science Publishers, New York (1972).

[22] J. M. Seitzman, "Quantitative applications of fluorescence imaging in combustion," Ph.D. Thesis, Stanford University, Palo Alto, CA (1991).

[23] S. D. Pack, M. W. Renfro, G. B. King, and N. M. Laurendeau, "Laser-induced fluorescence triple-integration method applied to hydroxyl concentration and fluorescence lifetime measurements," *Combust. Sci. Technol.* 140, 405–425 (1998).

[24] C. S. Cooper and N. M. Laurendeau, "Laser-induced fluorescence measurements in lean direct-injection spray flames: technique development and application," *Meas. Sci. Technol.* 11, 902–911 (2000).

[25] M. W. Renfro, "Time-series measurements of laser-induced OH and CH fluorescence in laminar and turbulent flames," M.S. Thesis, Purdue University, West Lafayette, IN (1997).

[26] P. H. Paul, J. A. Gray, J. L. Durant Jr., and J. W. Thoman Jr., "Collisional quenching corrections for laser-induced fluorescence measurements of NO A^{2}S$^+$," *AIAA J.* 32, 1670–1675 (1994).

[27] P. P. Yaney, T. P. Grayson, and J. W. Parish, "Measurements of temporal and spatial scales in gas flowfields using a two-pulsed-laser correlation scheme," *Proc. Combust. Inst.* 23, 1877–1883 (1990).

Composite Membrane for High Temperature Operation of Fuel Cells Based on Polymer Electrolytes

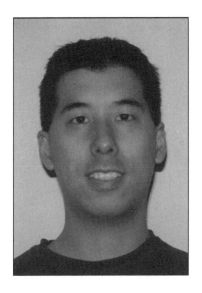

Christopher Yang

Center for Energy and Environmental Studies
Department of Mechanical and Aerospace Engineering
Princeton University
Olden Street
Princeton, NJ 08544
Research Advisors: Dr. Robert Socolow and Dr. Supramaniam Srinivasan

ABSTRACT

    Improving the CO tolerance of Proton Exchange Membrane Fuel Cells and achieving higher power densities in Direct Methanol Fuel Cells (DMFC's) are two research goals which need to be addressed in order to help bring these technologies to commercial market. One method for achieving both of these goals is to operate these fuel cells at higher temperatures. Traditional Nafion membranes are unable to operate at high temperature because of their dependence on water for proton conduction. Composite membranes based upon Nafion 115 and zirconium hydrogen phosphate can improve the performance of PEMFCs and DMFCs operating at high temperatures, by increasing the water retention properties. These membranes were synthesized and tested in a variety of temperature and humidification conditions to show improved water retention and improved fuel cell performance. High temperature operation of PEMFCs up to 130°C has been shown which should significantly increase CO tolerance of these fuel cells. Higher power densities have been demonstrated for DMFCs up to 150°C, as well as reduced humidification requirements. In addition to a performance evaluation, chemical and physical characterization of these membranes was performed. Based upon these experimental results, a theoretical analysis of these composite membranes and their water retention behavior is also introduced.

INTRODUCTION AND BACKGROUND

Fuel cells are electrochemical devices that are able to convert the chemical energy of a fuel (most often H_2) directly into direct current electrical energy with fairly high efficiency. In addition to improved efficiency, fuel cells running on hydrogen or methanol produce very low quantities of primary pollutants (such as NOx, SOx and particulates). Because of these benefits, the fuel cell has recently received a great deal of attention from the research and development community and the popular media as a dream technology which can be used in applications ranging from cell phones to automobiles to powerplants [e.g. 1–4]. The proton exchange membrane fuel cell or PEMFC, which is based on a polymer electrolyte, is the type of fuel cell that is most likely to be used in many of these applications. The direct methanol fuel cell or DMFC is based on the same membrane technology but uses a liquid methanol feed at the anode. Because the electrolyte is a strong polymer with high conductivity, these two fuel cells are fairly stable and are able to operate at low temperatures.

However, before these fuel cells can be commonly found in a wide range of applications, there are still a number of technical and economic issues which should be settled. One major technical issue is the issue of the type of fuel used and how that fuel can be delivered to the fuel cell. The PEMFC and DMFC take different approaches to the issue of the fuel supply. Hydrogen is the best fuel for PEMFCs and the use of reformed hydrogen is an important issue. H_2 needs to produced from other sources such as fossil fuels, alcohols, or water. The most likely starting materials for hydrogen production are carbon based fuels (natural gas, coal, gasoline or alcohols). In the process of producing a mixture of H_2 gas and CO_2, small amounts of CO will also be produced (from several ppm to around 5%) which poisons the platinum electrocatalyst at the anode. Several approaches to minimizing the effects of CO poisoning are being investigated, including the use of alloy catalysts, selective oxidation and operation of the fuel cell at elevated temperature. The latter is the focus of this research. By operation of a PEM fuel cell at temperatures of around 120–150°C, CO adsorption onto the platinum electrocatalyst is reduced and the tolerance level of the fuel cell increases dramatically from around 10 ppm at 80°C to around 1000 ppm at 130°C. The improvement of CO tolerance at increased temperatures could lead to the reduction in the complexity of the fuel processing system. Other benefits to high temperature operation include better heat and water management of PEMFCs and improved reaction kinetics for direct methanol fuel cells (DMFCs). Direct methanol fuel cells use the same type of membrane and similar Pt based electro-catalysts as PEMFCs. DMFCs have the benefit of using a liquid fuel directly and eliminates the need for a separate fuel processor. However, DMFCs are limited in power density because of slow electro-oxidation of methanol to CO2 at the anode. By increasing the fuel cell temperature, the kinetics of that reaction would be increased and higher power densities and lower catalyst loading could be realized.

Many researchers are working on operating PEMFCs and DMFCs at high temperature, but reaching high temperatures has been a difficult challenge because of the dependence of these polymer membranes on water. The state-of-the-art membranes for these fuel cells are perfluorinated sulfonic acid membranes such as Nafion® and Aciplex®. These membranes have a multi-phase structure [5]: one phase contains the hydrophobic fluorocarbon backbone and another, the hydrophilic ionic acidic region. Formation of these self assembled films either in the typical melt extruded or recast varieties creates a phase-separated ion-cluster network of sulfonic acid containing pores and channels which provide a continuous pathway through the membrane. Water is essential for proton conductivity within the membranes because it allows the hydrogen ions (H^+) of the sulfonic acid to dissociate from the acid and become solvated (H_3O^+). These hydrated ions are highly mobile and can move in the presence of an electric field. Under typical low temperature operating conditions, care must be taken to ensure that these membranes do not dry. This involves humidifying one or both of reactant gas streams or relying on product water to maintain membrane hydration. Use of these membranes at higher temperatures leads to evaporation of water from the membranes, a shrinking of the pores, and a loss of membrane ionic conductivity. Changes in the conductivity by over several orders of magnitude have been correlated with altering the water content of the membrane. [6,7] High protonic conductivity in the membrane is crucial to achieving high power density. Low membrane hydration leads to large ohmic losses which lowers operating voltage, power, and efficiency at a given current. As a result, reductions in proton conductivity at higher temperatures could potentially offset any performance benefits that would arise from higher CO tolerance in PEMFCs. Achieving conductivity and performance for the high temperature membrane that is comparable to state-of-the-art low temperature (80°C) membranes running on H_2 is a goal of this research. Another goal is to increase the operating power density of DMFCs by operation at higher temperatures.

Approaches to High Temperature Operation

Several approaches have been undertaken in this study to try to prevent the loss of water from the ionic regions (pores) of the membrane thereby maintaining the conductivity of the ionomer membranes at temperatures above 100°C. One of these is to incorporate a hydrophilic, inorganic material into the perfluorinated ionomer membrane. The efficacy of these hydrophilic additives has been demonstrated in the case of heteropolyacids in Nafion [8]. The rationale for this approach is that water molecules will be strongly hydrogen bonded to ions or dipoles in the inorganic material forming layers of hydration. This strongly held water will be less likely to evaporate. The second approach is the use of a non-aqueous, low volatility solvent to replace water as the proton acceptor within the perfluorinated ionomer membrane. Replacing water as the primary proton carrier has been demonstrated in membranes with phosphoric acid, imidazole, butyl methyl imidazolium triflate, and butyl methyl imidazolium tetrafluoroborate [9–12].

Another method is to use a solid state protonic conductor whose conduction mechanism occurs in the absence of water. This mode of solid state ion conduction is quite commonly used at much higher temperatures for oxide ion conduction in solid oxide fuel cells, though this hasn't been demonstrated in a fuel cell at low temperatures.

Several types of composite membranes were prepared and characterized for operation of PEMFCs at above 100°C. Though a preliminary attempt was made for each of these three approaches, the bulk of the work has been on the first approach: to modify these high conductivity ionomer membranes with hydrophilic materials to improve their water retention properties and maintain conductivity at elevated temperatures.

Previous work on Nafion/heteropolyacid composite membranes [8] showed good PEMFC performance at 115°C. However, while the acid is able to reduce the volatility of the membrane water and thus maintain sufficient membrane hydration at elevated temperatures, it has been shown that over time the liquid acid leaches from the membrane reducing the long-term effectiveness of the composite membrane. To overcome this problem, solid materials that can be immobilized in the membrane are considered. These include particulate and polymeric solids, such as silicon oxide, sulfated zirconia, and zirconium phosphate. DMFCs have been operated at temperatures up to about 145°C with a modified recast Nafion/SiO_2 membrane [13].

The experimental work focused on the preparation, characterization and testing of Nafion® 115 membranes modified by the addition of zirconium phosphate, $Zr(HPO_4)_2$. The synthesis and experimental results of these zirconium phosphate membranes provides new insights into the structural and chemical modifications to proton conduction and the expected temperature limitations of similar composite membranes.

EXPERIMENTAL WORK AND RESULTS

Membrane Electrode Assembly (MEA) Preparation

Standard Nafion 115 membranes (1100 g/mol SO_3^-, 125 mm thickness) and solubilized Nafion solution (equivalent weight 1100 g/mol SO_3^-) from Dupont was used in the preparation of the membrane electrode assemblies. Composite Nafion/zirconium phosphate ($Zr(HPO_4)_2$) membranes were prepared based upon a procedure of Grot and Rajendran [14] by swelling pre-treated Nafion 115 in a 1:1 volume mixture of methanol and water at 80°C and placing the swollen membranes in 1 M $ZrOCl_2$ (Aldrich) for 2 hours at 80°C to introduce the zirconium into the membrane (ion exchange of Zr^{4+} ions for H^+ ions) and then immersed in 1 M H_3PO_4 (80°C) overnight. The membrane was rinsed in boiling water for several hours to remove excess acid, dried, weighed, and the thickness measured. The form of zirconium phosphate in the membrane is believed to be a crystalline form called α-zirconium phosphate ($Zr(HPO_4)_2$) [15]. These membranes increase in weight by about 25% and thickness by about 40% from 5 mils to 7 mils as compared with the Nafion 115.

For hydrogen fuel cells, membrane-electrode assemblies (MEAs) were prepared using commercial ELAT 5 cm^2 electrodes (20% Pt on carbon, 0.4 mg Pt/cm^2 from E-TEK), impregnated with 0.6 mg/cm^2 of solubilized Nafion solution (dry weight). These electrodes were hot-pressed onto both sides of the membranes for 2 min at 130°C and 1 metric ton force. MEAs for use in DMFCs were made using electrodes prepared at CNR-TAE. The anode electrocatalyst for methanol oxidation is 1:1 Pt/Ru with a loading of 2.3 mg Pt/cm^2. The cathode electrocatalyst for oxygen reduction was 2.5 mg Pt/cm^2 of 20% Pt/C. These catalyst layers included 33% Nafion and supported on carbon cloth and a teflonized diffusion layer. The active area of these electrodes was also 5 cm^2. The MEA was prepared by hot-pressing the electrodes onto the membrane at 130°C at 50 bar pressure.

Membrane Characterization

X-Ray Diffraction (XRD) analysis was performed on the Nafion 115/zirconium phosphate membranes with several levels of membrane hydration (shown in Figure 1): (i) wet, hydrated membrane; (ii) partially hydrated membrane and (iii) a thoroughly dried membrane.

A dramatic difference exists between the three different scans due to the membrane hydration. The hydrated membrane has only one small peak (2.64Å) whereas the well dried membrane has sharp well defined peaks. Analysis of the crystalline peaks on the composite membrane show 2 peaks that are attributed to Nafion (5.2 Å and 2.3 Å) and several others that are attributed to the presence of zirconium phosphate (4.5, 3.73, 2.64, 1.7 Å). Some of these peaks match up with typical peaks found in various forms of zirconium phosphate (4.46, 3.54Å). The analysis is not definitive as to the form of zirconium phosphate found in each state of hydration. Most likely, several forms of zirconium phosphate exist in the membrane.

The XRD was used to calculate the particle size of the zirconium phosphate based on the location and width of the peaks. The zirconium particles were calculated to be 10 to 11 ±1 nm in size. The pore sizes in Nafion

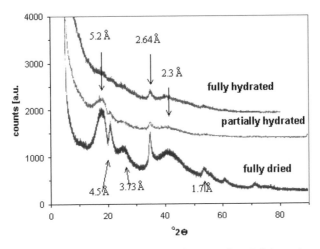

Figure 1. X-ray diffraction patterns for three Nafion®/zirconium phosphate membranes in varying states of hydration.

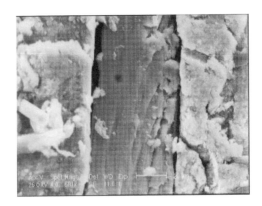

Figure 2. Scanning electron micrograph of a cross section of the MEA after operation in DMFC. Cathode, left; anode, right.

are on this order. Particles much smaller than the pores of Nafion could be lost from the membrane during preparation and fuel cell operation. Dried Nafion/zirconium phosphate membranes are 20–30% thicker than the starting Nafion 115 indicating significant swelling of the ionomer pores with zirconium phosphate.

The SEM analysis (shown in Figure 2) was performed on an MEA after 2 full days of operation in the DMFC. It showed that in the dried membrane after operation in the DMFC is about 100 μm thick, as compared to 125 μm for the starting Nafion 115 and 160 μm for the dried Nafion/ZrP membrane. EDAX analysis determined the presence of the following elements in the membrane-electrode assembly: C, S, Pt, Ru, F, Zr, P, O.

In order to determine the distribution of zirconium phosphate in the membrane, the electron beam was localized to a small area and measurements of the F:Zr ratio was measured at several places along the cross section of the membrane. A set of measurements were taken on a section of membrane located within the electrodes. In this case, a fairly uniform distribution of zirconium was found across most of the membrane. However, a reduction in the amount of Zr at the anode side as determined by a semi-quantitative comparison of the Zr to F ratio. The anode was identified by the presence of Ru in the catalyst layer of the MEA.

There is a reduction in the amount of Zr very close to the anode. This reduction is not seen at a distance of around 10 mm from the anodic membrane/electrode interface. It is believed that due to the presence of high temperature methanol/water mixture at the anode, swelling of the membrane can occur to the extent that some of the zirconium phosphate can be physically removed from the membrane. The effect is larger (though not quantified) at the location where the membrane was not surrounded by the electrodes. In this location, the membrane can swell much more than when confined by the electrodes or gasket and the loss of Zr appears much more signficant.

A long term study would be necessary to see whether this effect is localized merely to the first few microns of the anode side or if, over time, zirconium phosphate would continue to be removed from the membrane. If this effect remains localized very close to the anode, it would not have a significant effect on the operation of the fuel cell as there is more than sufficient membrane hydration at the anode due to the liquid feed.

Performance Evaluation

Conductivity measurements were made on the bare composite membrane, rather than the MEA external to the fuel cell environment. The membrane was held in the longitudinal position between small graphite electrodes and this assembly kept at 100% relative humidity at various temperatures. The measurements were made using a two probe method on a Princeton Applied Research potentiostat/galvanostat Model 273 and Princeton Applied Research lock-in amplifier Model 5210, connected to a PC running Electrochemical Impedance Software (EIS). In addition to external conductivity measurements, the current interrupt technique was used to make resistance measurements during fuel cell operation [16,17].

Conductivity of the Nafion/zirconium phosphate membrane was found to be about 15% lower at any given temperature than that of the unmodified Nafion 115 membrane when both are fully hydrated, indicating that

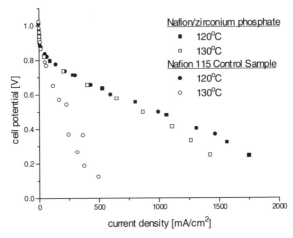

Figure 3. Comparison of polarization data at 120°C and 130°C for Nafion/zirconium phosphate and unmodified Nafion control sample at 3 atm system pressure in H_2/O_2 and 130°C vapor and reactant gas feed.

the displacement of some membrane water by the zirconium phosphate has a slight effect on the conductivity.

Testing of the membrane-electrode assemblies in a single cell allows for evaluation of the membrane and electrode performance under fuel cell operating conditions. The prepared MEAs were placed in 5 cm^2 single cell test fixture and the gas pressure, flow rate, temperature, humidification and single cell temperature were controlled using a Globetech™ fuel cell humidification system. PEMFC testing was carried out with pure H_2 and O_2 or air, and the level of humidification of these reactant gases was controlled by passing the gas streams through bottles of water of controlled temperature. For DMFC operation, a mixture of methanol and water was introduced to the anode side via a peristaltic pump after being preheated while the cathode was fed humidified or dry O_2 or air. The temperature, humidification, and pressure conditions were varied and current-potential (i-V) curves were generated using an Amrel Fuel Cell Electronic Load (FEL-60). Current density vs. potential (i-V) behavior was measured on these samples at a variety of temperature and humidification conditions.

Proton Exchange Membrane Fuel Cells (PEMFC)

Figure 3 shows a comparison between the performance of the Nafion/zirconium phosphate membranes and that of the unmodified Nafion 115. A dramatic difference exists at 130°C though the curves are identical at 120°C. Several factors affect the performance of the fuel cell as the temperature of the fuel cell and humidification bottles is varied. Conductivity of a fully hydrated membrane is a function of temperature as increased temperatures will allow for faster proton transport. Increased humidification bottle temperature is necessary to ensure adequate humidity in the gas stream to maintain membrane hydration, but when system pressure is held constant, increasing the concentration of water vapor reduces the partial pressure of the reactant gas and tends to lower performance.

At a total system pressure of 3 atm, a transition in performance occurs between 120 and 130°C. For full membrane hydration, it is necessary that the humidification bottle temperatures are higher than the fuel cell temperature. Because the vapor pressure of water is 2.6 atm at 130°C, it is not possible to increase the humidification temperature past 130°C without severely restricting the reactant gas partial pressure. Thus, for unmodified Nafion, there is a large drop when operating at 130°C as compared with 120°C. Water evaporates from the membrane and the pores shrink increasing the cell resistance. The Nafion/zirconium phosphate membrane is prepared in such a way as to promote a continuous structure which occupies the ion-cluster domain of the Nafion membranes. The zirconium phosphate is introduced homogeneously into the pores of the Nafion and subsequently improve water retention and conductivity throughout. Thus, at 130°C there is only a slight increase in the resistance of the membrane, indicating little loss of water, as compared with 120°C and at a typical operating voltage (0.6–0.7 V) there is almost no difference. Operation of the zirconium phosphate membrane has significant advantages over the Nafion membrane at 130°C.

Direct Methanol Fuel Cell (DMFC)

With the zirconium phosphate composite membrane, there was an increase in performance moving from higher humidification conditions to dry conditions at the cathode. As is typical with DMFCs using unmodified Nafion, increasing the cell temperature leads to higher current-voltage performance, higher maximum power density and higher power density at 0.5 V. However, unlike unmodified Nafion, the DMFC with the composite membrane had a maximum power at 150°C as compared to 130°C resulting in a greater maximum power density (see Figiure 4). The performance drops off above these temperatures. Under all humidification conditions, there is very little difference between the current potential curves for the temperatures from 120°C to 150°C at low to medium current density (0–600 mA cm^{-2} or potentials down to about 0.4 V). At high current density (around 1000 mA cm^{-2} and higher) there is much more of a difference between these temperatures with improved performance in the mass transport region at higher temperature and consequently leading to higher maximum power density.

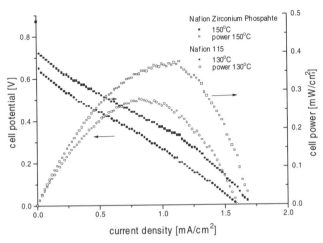

Figure 4. Comparison of polarization and power density data for Nafion/zirconium phosphate and unmodified Nafion control sample at 3 atm system pressure in 2M methanol/O_2. The Nafion/zirconium phosphate is at 150°C with 85°C anode methanol feed and unhumidified 85°C cathode feed, while the unmodified Nafion 115 has a 140°C anode methanol feed and 100°C humidified cathode feed.

There is a significant improvement in operating with dry oxygen especially at lower temperatures (90–110°C). In air, the effect is not as dramatic, but the same trend is seen. The reduction in cathode humidification leads to an increased performance most likely because of the reduction in flooding at the cathode side. The ability of the membrane to retain water is enhanced by the addition of the zirconium phosphate and as a result the need for cathode humidification is greatly reduced. This is in contrast to the inability of bare Nafion to operate at low temperature (85°C) cathode humidification. The composite membrane has reduced evaporation from the cathode side and consequently adequate membrane hydration due to the movement of water from the anode side to the cathode side and water production at the cathode. In the three conditions tested, the vapor pressure of water in the cathode gas stream was 1 atm, 0.6 atm, and 0 atm respectively. The performance also increased due to the increased O_2 reactant pressure. Another contribution to the improved performance is that there may be less methanol adsorption on the cathode catalyst because it is drier. Methanol adsorption is more pronounced in the presence of H₂O.

At very high temperatures, the operation of dry cathode gas is equal to or lower than that of the low temperature humidification. The rate of evaporation at these temperatures is faster than the supply of water via transport from the anode or production at the cathode and the membrane is not as well hydrated as with the humidified cathode gas.

Theoretical Analysis

The relative temperatures between the fuel cell and humidification system determines the relative humidity in the gas streams being fed to the fuel cell. This humidification is required to prevent dehydration of the membrane and loss of conductivity. The change in the chemical properties of these composite membranes as compared to the unmodified Nafion membranes can have an effect on the conditions necessary to maintain adequate hydration. Chemical thermodynamics can explain the effects that have been observed in the experimental results. The chemical potential of a liquid phase is essentially constant while the chemical potential of the vapor will vary with the vapor pressure.

$$\mu(g) = \mu(l) + RT \ln\left(\frac{P_{vap}}{P_{sat}}\right) \ . \tag{1}$$

If RH < 100%, the vapor pressure is less than the saturation vapor pressure ($P_{vap} < P_{sat}$) so that the vapor phase chemical potential is lower than that of the liquid phase and net evaporation will occur. The greater the deviation from 100% RH, the greater the difference between the two chemical potentials and the larger the driving force towards equilibrium. While the situation for the equilibrium of water in the membrane has similarities to the case of pure water, there are some key differences. Water does not exist as a pure liquid in a Nafion membrane. Sulfonic acid groups in the membrane can hydrogen bond with the water and this interaction serves to reduce the vapor pressure. A perfluorosulfonic acid membrane is different from an aqueous acid (e.g. 1 M H_2SO_4) because in the former, the acid groups are relatively fixed while in the latter they are mobile.

According to some previous studies [e.g. 6,7,18,19], the ratio of water to sulfonic acid ($\lambda = N_{H_2O}/N_{SO_3H}$) is between eighteen to twenty-two for a fully hydrated Nafion membrane of equivalent weight 1100. The water within the pores of the membrane are of three different types—there are water molecules that directly form hydrogen bonds with the electronegative oxygen atoms of the sulfonic acid; there is a secondary hydration sheath with weak hydrogen bonding; and rest of the water molecules that are in the membrane are not associated with these acid groups and are essentially like bulk water. In the membrane, the chemical potential of all the water molecules is not equal. The water molecules of the primary hydration sheath have the lowest values because of the strong ion dipole attraction. There is also a negative enthalpy of hydration for the hydrogen ion (H^+) from the sulfonic acid group which contributes to the association between water and acid group. There is a limit to the number of water that can form the primary hydration sheath. As that phase is completely filled, water would then occupy the phase with the next lowest chemical potential. That phase is the secondary hydration sheath because of the indirect interactions with the acid groups. Finally, the remaining bulk water in the membrane has the highest chemical potential, similar to that of pure water. Gottesfeld and Zawodzinski have proposed that this bulk water has an even higher chemical potential than pure liquid water [20].

Thus, the bulk water is the most likely to transfer to the vapor phase outside the membrane when the humidification conditions are unfavorable (RH<100%). The loss of the majority of the bulk water leads to a reduction in the degree of swelling in the membrane. As pore sizes in the membrane are reduced [5] resulting from this water loss, the membrane resistance increases because of lower water mobility in the membrane. Zawodzinski et al. have reported increased diffusion coefficient of water as the water content, λ, increases [19]. Also, the speed of proton hopping via the Grotthus mechanism in the remaining water could be reduced from that in bulk water due to the configuration of the waters around the sulfonic acid ions.

The effects of these modified, composite membranes can be seen by their effect on the membrane water. Their presence also leads to a loss of bulk water in the membrane, by displacement. While this does not increase conductivity at favorable hydration conditions (again, the mobility and proton hopping could be slightly reduced because of the lack of bulk waters), the introduced materials serve to add even more sites for hydrogen bonding and as the foundation for more hydration layers. This reduction in bulk water and increase in tightly and loosely bound water, and leads to a reduction in overall chemical potential of water in the membrane. As a result, while conductivity of these composite membranes is slightly lower because of the loss in bulk water, the presence of these materials can prevent water loss in lower humidity conditions than unmodified Nafion membranes.

The effects of these modified membranes on the vapor pressure can be accounted for in their effect on the free energy of vaporization. Hydrogen bonding with sulfonic acid groups or with modified membrane materials such as zirconium phosphate will increase the free energy of vaporization, according to the equation:

$$\ln\frac{P_2}{P_1} = \frac{-(\Delta G^\circ_{vap,2} - \Delta G^\circ_{vap,1})}{RT} = \frac{\mu_{l,2} - \mu_{l,1}}{RT} \quad , \tag{2}$$

where P_1 is the saturation vapor pressure at a given temperature for pure water and P_2 is the vapor pressure of water in the ionomer membrane, ΔG°_{vap} is the standard free energy for the vaporization of water, this equation gives the minimum ratio of vapor pressures (i.e. the relative humidity) that will still give thermodynamic membrane hydration. The term in the numerator on the right side of the equation are those enhancements to the free energy of vaporization that result from the factors mentioned. If the liquid phase chemical potential in case 2 is lower than for pure water, the vapor pressure of the vapor phase above 2 will be lower as well.

In addition to increasing sites for hydrogen bonding, another means of retaining water in a dehydrating environment involves use of microporous materials. A reduction in pore and channel sizes in the Nafion membrane could further reduce vapor pressure (and chemical potential) of water in the membrane by surface tension and capillary action effects. Assuming a channel of uniform diameter, the reduction in vapor pressure can be related characteristic size of the pore channels, given by the equation

$$P_{vap} = P_{vap}^{*} \, e^{\left(\frac{-2\gamma V_{m}}{rRT} \right)} \qquad , \qquad (3)$$

where P_{vap}^{*} is the vapor pressure in the absence of capillary action (bulk water), V_{m} is the molar volume of water, γ is the surface tension of liquid water and r is the radius of the pore.

A distribution in sizes of the pores and channels exists in a Nafion film (~10–1000 A) , and are essentially distributed randomly throughout the membrane. Under certain conditions, an equilibrium could exist whereby some of the smaller pores are filled whereas the larger ones are only partially full of water. Because of the need for a continuous pathway through the membrane for ionic conduction, this phenomenon leads to an increase in overall membrane resistance.

In addition to stronger retention of water in the membrane, another benefit to these composite membranes with hydrophilic additive is that these additives are introduced into the pores and channels of the Nafion membrane which reduces their effective radii. Further, the impregnation procedure of zirconium phosphate leads to the largest pores being filled with more material than the smaller ones, which reduces the range in pore free space. This would lead to a further reduction in the equilibrium vapor pressure and an increase in the amount of water hydrating the membrane at high temperatures.

Each of these effects which can enhance the membrane hydration by reducing the chemical potential of water in the membrane can also influence the kinetics of evaporation. The increased heat of vaporization (activation energy of reaction) leads to a reduction in evaporation rate based upon Arrhenius-type kinetics. This effect may play an important role when water is being supplied as in the anode feed of the DMFC. Even if the equilibrium system conditions would imply a dehydrated membrane, the rate of evaporation can be slower than the rate of water addition externally thereby maintaining membrane hydration. Even operating the fuel cell at high current density can produce water at a rate which slows or halts the attainment of equilibrium.

CONCLUSIONS

This paper focused on two aspects of high temperature polymer electrolyte membranes for PEMFCs and DMFCs, an experimental section detailing work with Nafion/zirconium phosphate composite membranes to achieving higher temperature operation and a subsequent theoretical framework for understanding the thermodynamics of water balance in the membrane. The experimental results show good performance for the zirconium phosphate membranes. These membranes have been shown to show good performance at 130°C and 3 bar pressure, conditions at which an unmodified Nafion membrane in a PEMFC performs poorly. In DMFCs, the composite membranes have shown an ability to operate at higher temperatures and to run under conditions of unhumidified cathode gas feed.

High temperature operation of fuel cell membranes that rely on water is a difficult proposition when operating with hydrogen, because of the exponential increase in the vapor pressure of water with temperature. Operation of a high temperature methanol fuel cell is an easier proposition because of the high quantity of water at the anode. The concepts developed are useful for addressing the challenges that arise in trying to operate a fuel cell at elevated temperatures and for determining the limitations and the extent to which a given approach can reduce the humidity requirements for successful operation.

High temperature membranes offers the promise of many benefits including higher CO tolerance at the anode electrocatalyst when utilizing processed carbonaceous fuels, higher power due to increased reaction kinetics, better water and heat management, and better prospects for waste heat utilization for fuel processing or space and water heating. These advances and other engineering advances and fundamental changes in fuel cell technology can continue to improve the performance of PEMFCs and DMFCs. Fuel cells can have a large impact on our energy future, in terms of limiting the emissions of CO_2 and other pollutants, and increasing our energy independence. Fuel cell research can lead us closer to the day where clean energy conversion technologies can be a major component of our generation capacity, and our energy needs will have minimal impacts on the environment.

ACKNOWLEDGEMENTS

I would like to thank my departmental advisor and numerous faculty/staff members at Princeton helping to steer my research, Prof. Rob Socolow, Dr. Supramaniam Srinivasan, Dr. Joan Ogden, Prof. Andrew Bocarsly, and Prof. Jay Benziger. I would also like to acknowledge fellow researchers Kev Adjemian, Paola Costamagna, and Seung Jae Lee as well as other members of the Center for Energy and Environmental Studies. Researchers from the CNR-TAE Institute in Messina, Italy, Dr. Vincenzo Antonucci, Dr. Antonino Arico, Enzo Baglio, Pasquale Creti, Dr. Enza Passalaqua, and Irene Gatto are recognized for their help with work on DMFCs and PEMFCs. Finally I would like to thank the Link Foundation, the Department of Mechanical and Aerospace Engineering, as well as the Department of Energy CARAT program for support of this research.

REFERENCES

[1]   A.J. Appleby, "The Electrochemical Engine for Vehicles" *Scientific American,* 281 (1) 74–79 (1999).
[2]   A.C. Lloyd, "The Powerplant in your Basement" *Scientific American,* 281 (1) 80–86 (1999).
[3]   C.K. Dyer "Replacing the Battery in Portable Electronics" *Scientific American,* 281 (1) 88–93 (1999).
[4]   M.M. Steinbugler, R.H. Williams, "Beyond Combustion–Fuel cell cars for the 21st Century" *Forum for Applied Research and Public Policy,* 102–107 (1998).
[5]   T.D. Gierke and W.Y. Hsu, "The Cluster-Network Model of Ion Clustering" in *Perfluorinated Ionomer Membranes,* A. Eisenberg, H.L. Yeager, (eds), Washington DC, 283–307 (1982).
[6]   A.V. Anantaraman, C.L. Gardner, "Studies on Ion-Exchange Membranes. Part 1. Effect of Humidity on the Conductivity of Nafion" *J. Electroanal Chem.* 414, 115–120 (1996).
[7]   Y. Sone, P. Ekdunge, D. Simonsson, "Proton Conductivity of Nafion 117 as Measured by a Four-Electrode AC Impedance Method" *J. Electrochem. Soc.* 143 (4), 1254–1259 (1996).
[8]   S. Malhotra and R. Datta, "Membrane-Supported Nonvolatile Acidic Electrolytes Allow Higher Temperature Operation of Proton-Exchange Membrane Fuel Cells" *J. Electrochem. Soc.* 144 (2), L23–L26 (1997).
[9]   R. Savinell, E. Yeager, D. Tryk, U. Landau, J. Wainright, D. Weng, K. Lux, M. Litt and C. Rogers, "A Polymer Electrolyte for Operation at Temperatures up to 200°C" *J. Electrochem. Soc.* 141(4), L46–L48 (1994).
[10] K.D. Kreuer, A. Fuchs, M. Ise, M. Sapeth, J. Maier, "Imidazole and pyrazole-Based Proton Conducting Polymers and Liquids" *Electrochimica Acta,* 43 (10–11), 1281–1288 (1998).
[11] M. Doyle, S.K. Choi , G. Proulx, "High-temperature proton conducting membranes based on perfluorinated ionomer membrane-ionic liquid composites" *J. Electrochem. Soc,* 147 (1), 34–37 (2000).
[12] J.T. Wang, R.F. Savinell, J. Wainright, M. Litt and H. Yu, "A H$_2$/O2 Fuel Cell Using Acid Doped Polybenzimidazole as Polymer Electrolyte" *Electrochimica Acta,* 41 (2), 193–197 (1996).
[13] P.L. Antonucci, A.S. Arico, P. Creti, E. Ramunni and V. Antonucci, "Investigation of a direct methanol fuel cell based on a composite Nafion®-silica electrolyte for high temperature operation" *Solid State Ionics.* 125, 431–437 (1999).

[14] W.G. Grot, G. Rajendran, "Membranes containing inorganic fillers and membranes and electrode assemblies and electrochemical cells employing same" United States Patent. No. 5919583. (1999)

[15] G. Alberti, "Syntheses, Crystalline Structure and Ion-Exchange Properties of Insoluble Acid Salts of Tetravalent Metals and Their Salt Forms" *Accounts of Chemical Research*, 11, 163–170 (1978).

[16] M.C. Wintersgill, J.J. Fontanella, "Complex Impedance Measurements on Nafion" *Electrochimica Acta.* 43 (10–11), 1533–1538 (1998).

[17] J.J. Fontanella, M.G. McLin, M.C. Wintersgill, J.P. Calame, S.G. Greenbaum "Electrical Impedance Studies of Acid Form Nafion Membranes" *Solid State Ionics* 66, 1–4 (1993).

[18] K.D. Kreuer, "On the Development of Proton Conducting Materials for Technological Applications" *Solid State Ionics* 97, 1–15 (1997).

[19] T.A. Zawodzinski, Jr, C. Derouin, S. Radzinski, R.J. Sherman, V.T. Smith, T.E. Springer, S. Gottesfeld, "Water Uptake by and Transport Through Nafion 117 Membranes" *J. Electrochem. Soc.* 140 (4), 1041–1047 (1993).

[20] S. Gottesfeld, T.A. Zawodzinski, "Polymer Electrolyte Fuel Cells" in *Advances in Electrochemical Science and Engineering, Vol 5*, R.C. Alkize, H. Gerisher, D.M. Kolb, C.W. Tobias, eds. 197–301 (1997).

Optimal Design and Operation of Solid Oxide Fuel Cell Systems for Distributed Generation

Robert J. Braun

Solar Energy Laboratory
Department of Mechanical Engineering
University of Wisconsin-Madison
Madison, Wisconsin 53706
Research Advisors: Dr. Sandy A. Klein and Dr. Douglas T. Reindl

ABSTRACT

Optimal design and operation strategies for commercial- (200 kW) and small- (2 kW) scale solid oxide fuel cell power generators are analyzed. Results of the optimal design and simulation studies are divided into two parts. In the first part, selection of optimal fuel cell operating parameters for design of a commercial scale electric-only application are presented. The performance parameters studied are fuel utilization, operating cell voltage, and operating temperature. For a given cell area specific resistance, these parameters will dictate the power density and operating efficiency of the fuel cell stack module. The choice of the relevant performance parameters may maximize electric power or electric efficiency. Consideration is also given to the fuel cell operating envelope, which must be designed to accommodate off-design point operation for load-following situations. The results of the first part indicate that optimal cell voltage occurs at 0.7 volts, 89% fuel utilization, and a cell operating temperature of 800°C. Sensitivity analyses to both economic and operating performance parameters are also discussed. In the second part, the design methods from the commercial scale analysis are applied to a 2 kW residential fuel cell power generator with cogeneration in the form of domestic hot water. An annual simulation of the residential Solid Oxide Fuel Cell (SOFC) power cogenerator was carried out. Results indicate an annual cogeneration efficiency of 84% and a ten-year simple payback are possible. The serving of electric and hot water loads by the fuel cell system was facilitated by a 2–tank hot water storage system that enabled continuous heat recovery irrespective of thermal load. The low fuel cell electric capacity factor of 46% indicates that a fuel cell size of 2 kW, for the home in this analysis, may be too large.

INTRODUCTION

Background and Motivation

Electrochemical fuel cells have the potential to convert fuel directly to electricity and heat at efficiencies greater than any single conventional energy conversion technology. Their modular nature coupled with their ability to generate electricity in a clean and efficient fashion make them suitable for a wide variety of applications and markets. There are six different types of fuel cells that have received varying degrees of development attention. Presently, the 80°C proton exchange membrane (PEM) and the 700–1000°C solid oxide fuel cell (SOFC) have been identified as the likely fuel cell technologies that will capture the most significant market share [1,2]. As fuel cells are targeted for early commercialization in the residential (1–10 kW) and commercial (25–250 kW) end-use markets, system studies in these areas are of particular interest.

The advent of maturing fuel cell technologies represents an opportunity to achieve significant improvements in energy conversion efficiencies at many scales; thereby, simultaneously extending our finite resources and reducing "harmful" energy-related emissions to levels well below that of anticipated regulatory standards. Due to their modular nature, fuel cells have the potential to widely penetrate energy end-use market sectors. If fuel cells are applied at a large scale, substantial reductions in both national emissions and fuel consumption may be realized. Although fuel cells-themselves have been studied extensively, primarily from materials and electrochemical viewpoints, a considerable gap exists in the area of application techniques to maximize benefits of fuel cell units for both electrical energy generation and thermal energy utilization. With fuel cell commercialization just a few years away, much effort is still being expended on developing component level hardware, such as fuel cell stacks and fuel reformers, while relatively little work is being performed in systems-level research. In order to realize the high-energy conversion efficiencies offered by fuel cell devices, it is crucial that methodologies for system-level optimal design be developed to achieve the maximum overall system efficiency and cost effectiveness.

Objectives

It has often been stated that the characteristically high thermal-to-electric ratio of the SOFC makes them attractive for providing the thermal requirements of various end-use applications. The high-grade waste heat produced in a solid oxide fuel cell can be utilized for space heating, process steam, and/or domestic hot water demands. The utilizability of this waste heat can significantly impact system efficiency, economics, and environmental emissions. The type of heat recovery depends on the application requirements and the resulting cogenerative efficiency will depend on the design. A significant issue surrounding the use of fuel cells (and their efficiency) in residential appli-cations, is their ability to meet the highly non-coincident electric and thermal loads in either grid-connected or stand alone configurations. That is, in either base load operation or electric load-following conditions, electricity and/or heat may be available when it is not needed or vice-versa. Additionally, either higher or lower fuel efficiency and different proportions of electric and ther-mal output is derived from the fuel cell system depending on where the fuel cell stack is operated on its voltage-current characteristic. As a result, both the *system* design point and off-design point operating characteristics are depen-dent on (i) selection of optimal *fuel cell* design and operating point, (ii) heat recovery design, (iii) electric and thermal load management, and to a lesser degree (iv) the performance characteristics of auxiliary hardware, such as in-verters, pumps, compressors, controls, and external reformers (if any).

The successful design of fuel cell systems requires proper selection of cell stack operating conditions. Studies focusing on optimal operating point se-lection are emerging [3,4,5]. In addition to operating point selection for opti-mal design point operation, the fuel cell operating envelope must be designed to accommodate off-design point operation for load-following situations.

The objectives of the research can be summarized as (i) establish the optimal operating point for the purposes of system design through mini-mization of the system life cycle costs expressed as "cost-of-electricity" (COE), (ii) establish the sensitivity of economic and cell performance pa-rameters on operating point selection, and (iii) simulate an SOFC cogenerator in a residential application to gauge performance and to offer additional optimal design strategies. The studies are performed for an ad-vanced planar solid oxide fuel cell system employing indirect internal re-forming and anode gas recirculation.

This paper is divided into two parts. In the first part, optimal operating point studies on a commercial scale electric-only application are presented (cf. Braun et al., [6]). Sensitivity analyses to both economic and operating performance parameters are also discussed. In the second part, the meth-ods of the commercial scale analysis are applied to a smaller, residential fuel cell power generator with cogeneration in the form of domestic hot water. The performance of the residential power system is parameterized and employed for annual simulations.

SYSTEM DESCRIPTION

200 kWe Base Case System Design

A 200 kWe natural gas-fueled planar solid oxide fuel cell system provid-ing electrical power only was chosen for the operating point studies. A non-optimized reference design of the system with a nominal fuel cell op-

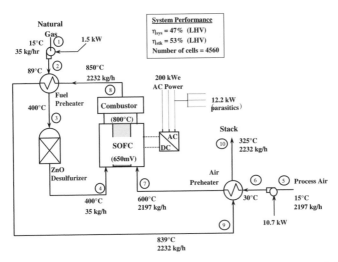

Figure 1. 200 kWe Planar SOFC Process Flow Diagram {adapted from [7]}.

erating temperature of 800°C is shown in Figure 1. Salient features of the basic system design include the use of thermally integrated indirect internal reforming, anode gas recirculation, two stage air and fuel preheat, reactant blowers, inverter, and a natural gas desulfurizer. The system under study is an advanced design in that heat removal from the stack is not strictly dependent on cathode air flow. The thermally integrated indirect internal reformer acts as an additional heat sink, thereby reducing cooling air requirements. The power module enclosure houses the fuel cell stack, fuel ejector, preheater, and reformer, and air preheater. A conceptual diagram of the stack processes is illustrated in Figure 2.

Natural gas enters the plant at station 1 and is pressurized to 2.13 bar. After preheat, the fuel enters the desulfurizer where sulfur levels are reduced below 0.1 ppm [7]. The desulfurized fuel enters the stack module at station 4. Air enters the plant at station 5 (420% theoretical) and is pressurized to 1.14 bars before entering the preheater. Air preheat to 600°C is accomplished through a cross-flow plate fin type heat exchanger with an effectiveness of 0.70. Inside the stack, the fuel is mixed and heated with a portion of the depleted anode gas such that the resultant fuel mixture has a steam-to-methane molar ratio of 1.6:1. After reforming, the hydrogen-rich gas stream enters the anode gas manifold and is electrochemically converted with a fuel utilization* of 85%, producing 226 kW of DC power which

---

*Fuel utilization is defined as the number of moles of hydrogen consumed divided by the molar amount of hydrogen supplied to the cell (including the hydrogen produced via reaction (2)).

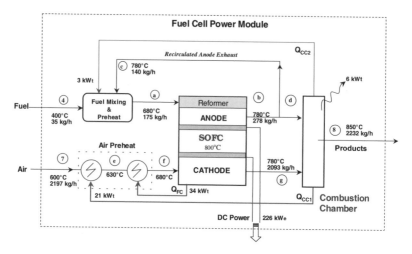

Figure 2. Conceptual Fuel Cell Stack Process Diagram {adapted from [28]}.

is inverted to AC with an efficiency of 94%. The net system electric efficiency is 47%. About half of the depleted fuel (by mass) that exits the anode compartment is catalytically combusted with the partially depleted cathode gas stream. Air entering the stack is distributed around the enclosure such that it is further heated from 600°C to 680°C before entering the cathode compartment. A temperature rise of 100°C occurs across the stack. After combustion, the product gases exit the stack at 850°C (station 8). The combustion gases provide the thermal energy for preheat of the process fuel and air streams before exiting the plant at 325°C.

Fuel Cell Stack

Direct electrochemical oxidation of methane is theoretically possible, but its use for near-term solid oxide cells employing zirconia-based electrolytes is not realistic. Methane is reformed prior to admittance into the cellstack to produce hydrogen according to reactions (1–3),

$$CH_4 + H_2O \leftrightarrow CO + 3H_2, \tag{1}$$
$$CO + H_2O \leftrightarrow CO_2 + H_2, \tag{2}$$
$$CH_4 + CO_2 \leftrightarrow 2CO + 2H_2, \tag{3}$$

where (3), while not independent of (1) and (2), is possible due to anode gas recirculation. Nearly 100% methane conversion is achieved in the reformer. The steam required for reforming is provided by anode gas recirculation. As the risk for carbon formation through side reactions is reduced by increasing the steam-to-carbon molar ratio (s/c) of the fuel mixture [8], a water vapor amount greater than the stoichiometric requirement of reaction (1) is required. A minimum value of 1.6 was chosen per Wagner and Froment [9] who indicate that on a thermodynamic basis, this value is sufficient to ensure no methane cracking. The fuel feedstock is heated to only 400°C before being admitted to the stack module. Additional heating to 680°C is accomplished by anode gas recirculation, and convective and radiative heat transfer from stack and combustor surfaces. The s/c ratio and the fuel utilization fix the amount of anode gas recirculation.

Cooling Air Requirements

Air-cooled solid oxide fuel cells typically operate with excess air several times above stoichiometric requirements due to the need for maintaining a small temperature rise (≤ 100°C) across the fuel cell stack. Bossel [10] reports theoretical air* requirements ranging between 300–1000% for SOFC systems. Typical values of 300–600% can be found in the literature [7,11–13]. A pinch temperature difference of 20°C was set between the solid cell material and the depleted reaction products. The maximum combustion products temperature leaving the stack module is constrained to 850°C by adjustment of the air flow to enable use of lower cost materials in the downstream fuel and air gas heaters. The temperature rise across the cell stack is set at 100°C and the temperature rise from reactant inlet (stations a and f) to combustion outlet is 170°C.

METHODOLOGY AND APPROACH

Simulation Program

A program to determine all the state point variables in the thermodynamic flowsheets detailed in Figures 1 and 2 was written using EES [14], a general purpose equation solver. Mass and energy balances were written for each component in the system. Performance characteristics, such as cell voltage-current curves, blower and compressor efficiencies, and heat exchanger effectivenesses were included in the analyses. Thermodynamic properties were computed using correlations provided by EES and the resultant system of nonlinear equations is then solved in the same program.

---

*Theoretical air is defined here as the amount of air necessary to oxidize the system fuel gas input.

Performance Characteristics

For atmospheric pressure operation, the voltage-current characteristic is approximated as a linear relationship with the Nernst voltage as the y-intercept and the area specific resistance as the slope. As the electrodes act as equipotential surfaces, the maximum potential difference possible is the lowest Nernst potential of the cell. The location of the lowest Nernst potential typically occurs at the cell outlet [15]. The anode gas consists of CO, CO_2, H_2O, H_2, and residual CH_4 species that rapidly reach equilibrium in the high temperature environment. An equilibrium calculation is performed at the cell outlet prior to input into the V-I characteristic.

The slope of the V-I characteristic is representative of the cell area specific resistance (ASR), which is temperature sensitive. Various sources for characterizing the temperature dependence of the ASR exist in the literature (cf. Ghosh et al., [16] and Minh [17]). The temperature dependence given by Chen et al., [3] is employed in the life cycle cost estimations of this analysis.

Lundberg [7] showed stack heat loss as a function of capacity to range from 1.0 to 3.5% of the fuel lower heating value for a tubular SOFC stack design. The stack heat loss parameter for this study is 1.5% of the system fuel energy input (LHV basis). Fuel cell stack and system efficiency are defined as,

$$\eta_{stk} = \frac{\text{DC Power Out}}{\dot{N}_{fuel,in} \cdot LHV_{fuel}} \quad \text{and} \quad \eta_{sys} = \frac{\text{Net AC Power Out}}{\dot{N}_{fuel,in} \cdot LHV_{fuel}}. \quad (4)$$

Table 1 details the important system hardware specifications employed, including that of the base case, range of parameter variation, and fixed variables.

Economic Considerations

For on-site distributed power generation, transmission and distribution costs do not factor into the cost of electricity. Annual operation and maintenance is estimated at 1% of the system capital cost. SOFC stacks are assumed to have an operational life of 5 years with a salvage value of 1/3 the original investment. The plant life is assumed to be 20 years with a capacity factor of 0.8. Fuel cost, unless noted otherwise, is $4/MMBtu. The cost of capital is 8% and the return on investment is 12% (before taxes), yielding a discount rate of 20%. The stack manufacturing costs given by Chen et al., [3] are employed in this analysis and are based on 200 MW/yr production levels. The SOFC costs are dependent on the materials employed and the cell component thicknesses. The economy of the cell stack is based on the use of Ni/Zr cermet anode (100 μm thick), yttria-stabilized zirconia electrolyte (5 μm), strontium-doped lanthanum manganite cathode (100 μm),

Table 1. Reference System Values and Parameter Ranges Under Study

Parameter	Ref. Value	Range Studied
Cell Temp. (°C)	800	700–1000
Cell Press. (atm)	~1	Fixed
Cell voltage (mV)	650	500–850
Power Output (kWe)*	200	may vary*
ASR (Ω cm^2)	.663	.278–1.01
Fuel Utilization (%)	85	60–95
Air Stoichs, S	4.2	2–6
S/C Ratio	1.6	Fixed
Cell stack ΔT (°C)	100	≤ 100
Module ΔT (°C)	250	200–450
Module Outlet Temp.	850	≤ 850
Fuel Compressor Effic.(polytropic)	70%	Fixed
Air Blower Effic.(static)	65%	Fixed
Fuel Preheater Effectiveness	.0035	Fixed
Air Preheater Effectiveness	0.70	variable
Inverter Efficiency	94%	Fixed

Table 2. System Capital Cost Data

Component	Unit Cost[1] ($/kWe)	Cost Exponent
Cell hardware[2]	$492/m^2	N/a
Inverter / controls	200	0
Air Preheater	106	0.75
Air Blower	43	1.0
Fuel Processing[3]	34	0.6
Aux. Hardware[4]	160	0

[1] 1999 US$
[2] 800°C cost shown. Cell costs are a decreasing function of operating temperature.
[3] Includes desulfurizer and catalyst, fuel preheat, and ejector.
[4] Includes startup burner and boiler, controls and instrumentation.

and either 1mm stainless steel (700-800°C), high alloy metal (900°C), or La Chromite (1000°C) for the interconnect depending on operating temperature. Cell electroactive area is 225 cm^2.

The balance-of-plant costs are adapted from Chen et al., [3], Lundberg [7], and Hsu [18]. Scaling of capital costs for the parametric optimization studies used the cost scaling techniques detailed in Boehm [19]. Cost scaling exponents were selected from Boehm, Perry et al., [20], and Peters et al., [21] and cost sensitivity analyses.

The variables considered in the operating point study include operating temperature, cell voltage, and fuel utilization. In addition to a technical study of operating variables, the sensitivity of minimum life cycle costs to economic parameters, such as fuel and cell stack cost are also presented.

OPERATING POINT ANALYSES

Two paths of study are available for the system analyses. The first path is one that fixes cell stack size (i.e., number of cells) and examines cost and performance changes due to variation of system operating parameters, such as operating cell voltage, fuel utilization, and power module inlet air temperature. The second path fixes power output while varying operating parameters. Both options require a cell performance characteristic and a general system process design (e.g., see Figure 1). Both options will also lead to the same optimum since the influence of cost-of-electricity is relatively independent of size. Riensche et al. [4] examined the first approach in some detail for a non-optimized planar SOFC system. In this study, only a brief examination of fixed stack size is presented before moving on to the method of fixed power output.

Fixed Stack Size

The effect of varying operating cell voltage on the cost-of-electricity and system electric efficiency is shown in Figure 3. At 650 mV operation, the reference COE is 6.3¢/kWh. The unit system capital cost* associated with an average 650mV cell voltage is 1100 $/kWe. As the operating voltage is increased (increasing fuel conversion efficiency), the cell-stack costs begin to increase at a rate greater than operating costs because lower current densities result and therefore larger cell areas are required. Continued increases in fuel efficiency cannot pay for increases in capital costs, which are dominated by the fuel cell stack, and the selling price of electricity must then be raised to compensate. The minimum cost occurs at a cell voltage of 700 mV with a corresponding system efficiency of 52%. The effect of fuel utilization on COE also exhibits an optimum characteristic near a utilization of 90% as shown in Figure 4. Increasing fuel utilization results in increasing capital costs due to the reduced average current density; but this effect is offset by decreasing fuel costs.

The model developed for a variation in the fuel utilization parameter shows some sensitivity to fuel reforming calculation methodology. In a first approach, the results of which Figure 4 illustrate, the conversion of methane and carbon monoxide via Equations (1) and (2) is assumed to be 100% (in the presence of recirculated gases). Use of this method suggests a slight over-prediction of maximum theoretical voltage and optimal fuel utilization. In a second method, the equilibrium composition of the reformate is calculated prior to entry into the anode. This approach indicates that optimal fuel utilization is lower than that depicted in Figure 4. The location of optimal cell voltage and cell temperature are unaffected by the choice of methodology.

---

*The unit system capital cost does not include installation, transportation, or contingency fees.

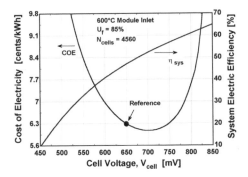

Figure 3. Variation of Cell Voltage.

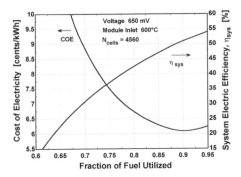

Figure 4. Variation of Fuel Utilized.

The impact of lowering the excess air requirement, by increasing the allowable temperature rise across the fuel cell power module (from station 7 to 8 of Figure 1), on the COE is reduced by the thermal management scheme of the internally reforming stack design. A 50°C decrease in inlet air temperature causes a reduction of only 17% in cooling air requirements and thus, only a small decline in COE. This result is shown in Figure 5 as a function of COE normalized to the reference design case of 6.3¢/kWh. Assuming that a 50°C decrease in inlet air temperature is permissible, moving to the optimal parameters suggested by Figures 4 and 5 (700mV, U_f=89%) results in a decrease of nearly 5% in COE and an increase in system efficiency from 47% to 55%. Power output, however, is reduced by 22%.

Fixed Power Output Analysis

In this analysis, a fixed power output of 200 kWe is specified and the variation of operating cell temperature, fuel utilization, cell voltage, and power module inlet air temperature are examined. The objective of the study

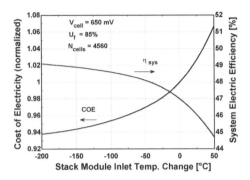

Figure 5. Variation of Inlet Temp.

is to determine optimal operating voltage, temperature, and fuel utilization for the system process design detailed in Figures 1 and 2. Implicit in
this analysis is that the stack V-I curve is not affected by changes in the
number of cells in the stack. Constraints on the system are the same as
listed in Table 1. As before, optimal operating parameters were selected on
the basis of minimum cost-of-electricity. However, due to uncertainty in
cost estimates, the COE results are normalized to the reference design COE
of 6.3¢/kWh.

Figure 6 shows the minimum COE as a function of both fuel utilization
and gas temperature rise across the power module for 200 kWe power output and a cell operating temperature of 800°C. In this figure, the optimal
operating cell voltage is determined at each specified value of fuel utilization. The optimal fuel utilization is 90% (at a cell voltage of 670 mV) with a
system efficiency of nearly 53%. The cost-of-electricity registered only 3%
below the reference design case. Additionally, to accomplish a 200 kWe
power output at the optimal parameters, the required number of cells in-

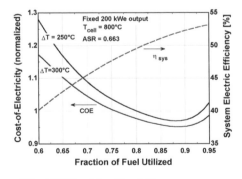

Figure 6. Variation of Fuel Utilized for Fixed Power.

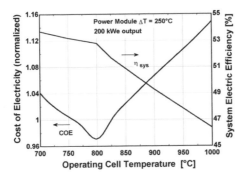

Figure 7. Optimal Cell Temperature.

creased by 17%. The effect of lowering the excess air required, by decreasing the inlet air temperature to the stack power module by 50°C, reduces the COE to 5% below the reference design.

The effect of operating cell temperature on COE and system efficiency is shown in Figure 7. The optimal cell voltage and fuel utilization at each cell operating temperature were employed. As cell temperature is reduced from 1000° to 800°C, interconnect material costs are reduced as ceramic materials may be replaced with stainless steel. However, while stack costs are reduced by a factor of 2.3, the area specific resistance increases by the same factor over this temperature range. Below 800°C, no stack cost savings are realized and performance continues to decrease, thus yielding the extrema shown in Figure 7. A similar result has been obtained by Chen et al., [3].

A breakdown of the various costs is shown in Figure 8. As the figure indicates, fuel cost is dominant, followed by cell stack and balance-of-plant (BOP). The bulk of BOP costs (64%) are attributable to the inverter and start-up equipment, instrumentation and controls. Aside from these costs,

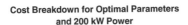

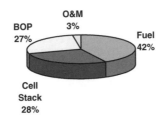

Figure 8. Breakdown of System Costs.

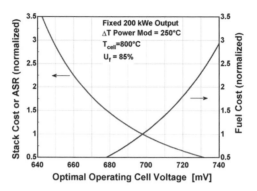

Figure 9. Effect of Stack and Fuel Costs.

the air preheater accounts for 16%, fuel processing 15%, and air blower 5% of the total BOP cost.

The effect of varying fuel cell stack cost, area specific resistance (ASR), and fuel cost is illustrated in Figure 9. The stack costs are linearly dependent on cell operating temperature. Since ASR is approximately linearly dependent in the same temperature range, the effect of normalized changes in stack cost and ASR on optimal operating voltage exhibit the same trend. The stack cost estimates utilized in this analysis are based on large manufacturing volumes (200 MW/yr). Early SOFC production units are likely to see higher costs, and as Figure 9 indicates, must operate at lower voltages for increased power density. The cost of natural gas can vary greatly depending on geographic location. Increase in fuel costs necessitates higher electricity prices to achieve adequate payback thereby boosting the required optimal operating cell voltage as Figure 9 shows.

Conclusions for Optimal Operating Point Analyses

To summarize, optimal operating voltage (0.7V), fuel utilization (89%), and operating temperature (800°C) were determined for both fixed size and fixed power analyses. System electric efficiencies as high as 55% are possible in these schemes. It is was also recognized that the increases in temperature rise across the fuel cell power module had little effect on COE. Economic sensitivities indicate decreasing optimal voltage for increases in stack costs and increasing cell voltages for increases in fuel cost.

The use of this methodology to select optimal operating points for *all* power capacities is valid, however, the location of the extrema will depend on the associated system component costs. As capital costs for smaller BOP components are larger (per kW), as in the residential fuel cell power system of Part II of this paper, the optimal operating point will be lowered.

Additionally, designing for a load-following situation presents additional challenges as fuel cell operating envelopes must be carefully selected to avoid operating regimes that adversely affect cell-stack components. For instance, the case of a high power output operating point translates into a high average current density (low cell voltage). Operation near maximum power can produce harmful effects in the form of local hot spots in the electrode that accelerate sintering of nickel particles and subsequent reduction of electroactive area [22].

SOFC SIMULATION IN A RESIDENTIAL APPLICATION

Residential SOFC System Description

A 2 kW rated SOFC system with a two-tank hot water storage configuration is shown in Figure 10. The unit capital costs ($/kW) for auxiliary hardware are larger for smaller fuel cell power systems. The primary cost increase is associated with the dc-to-ac inverter and instrumentation and controls. Unit cost increases are also anticipated for fuel processing, air blower, and air preheater. In terms of the fuel cell sub-system, the system shown in Figure 10 operates very similarly that of Figure 1, except that at the design point, the fuel cell power module air inlet temperature is 550°C and the exhaust gas temperature increased from 325° to 385°C. The optimal design parameters for this system (based on an electric-only analysis) were 670 mV, 75% fuel utilization, and 800°C cell temperature. Operating at the design conditions the system electric efficiency is 47.8% (LHV basis). The unit system capital cost for this design was estimated to be $1925/kWe. The total system capital cost, including the two hot water tanks, was $4100.

The conceptual system was designed for electric load-following and buffering of non-steady electrical demands by the grid. The two-tank hot water storage configuration allows for thermal buffering, that is, heat recovery during zero hot water draw situations. As Figure 10 illustrates, cold water from the main enters the first hot water tank at 10°C. Depending on the amount of heat recovery and the tank temperature, the cold water can be heated from 10° to 65°C. If the heated water leaves the first tank at a temperature lower than 60°C, a burner in the second tank accomplishes supplementary heating to the delivery temperature. A single node, lumped capacitance model is employed to simulate the hot water tank system. At rated power conditions, fuel cell waste heat gases enter the first hot water tank at 385°C where 1.6 kW of heat may be recovered before the gases exhaust the system at about 95°C (station 11). The corresponding thermal-to-electric power ratio is 0.8:1.

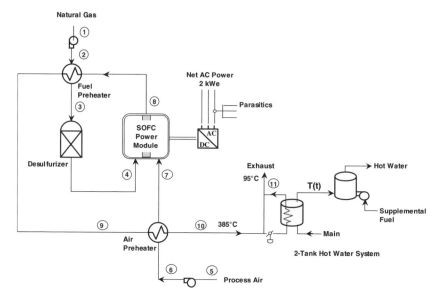

Figure 10. Residential SOFC System with Thermal Storage.

Operating Strategies

Electric load data for a winter day was obtained from GRI [23], and hot water heating data from Mutch [24]. Space heating and cooling data was generated for a 242 m² (2,500 ft²) home located in Madison, WI using a TRNSYS [25] Type56 model and typical meteorological year (TMY) weather data. The resultant electric, hot water, space heating, and cooling loads on "typical" winter and summer days are shown in Figure 11. In winter, the maximum electric load is approximately 1.5 kWe. The load data are represented by integrated hourly averages and therefore do not depict the shorter time scale peak power demands of 10 kW or more that characterize residential electricity consumption [26]. It is also noteworthy that the space heating thermal requirement can often be ten times greater than the electrical load. From this it can be seen that the use of residential fuel cell power systems to serve space-heating loads appears difficult to achieve without other system concepts. In contrast, the domestic hot water demand illustrates a better match between the magnitudes of thermal and electric loads.

On a "typical" summer day, the electric load increases due to vapor compression power requirements of air-conditioning systems. The peak power demand generally occurs in the afternoon hours and in this analysis, may range from 1.5–3.5 kW. Figure 11 indicates that although the coincidence of

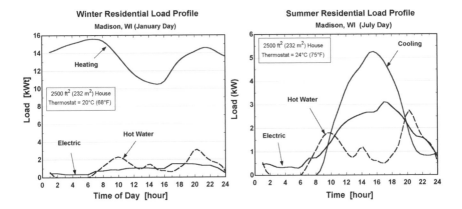

Figure 11. Winter and Summer Residential Load Profiles.

the electric and thermal loads is not well matched, electrical and thermal storage opportunities exist.

The fuel cell electric and thermal load-following performance for the winter and summer days is shown in Figure 12. The solid oxide fuel cell system was designed to handle a 4–to-1 electric turndown. Thus, for loads below 0.5 kW, the fuel cell was forced to shutdown. Shutdown periods last between three and four hours, and occur in the early morning hours. During times of high electric demand and low water draws, the first hot water storage tank was sized to ensure that tank temperatures would not exceed 90°C. Mixing cold water from the main could conceivably lower tank temperatures to accommodate the delivery temperature of 60°C. The main advantage of the 2–tank hot water configuration is that it effectively enables continuous heat recovery irrespective of the load as evidenced by the thermal load profile for the July day shown in Figure 12.

Annual Simulation Results

The utility electricity price for this analysis was 7 ¢/kWh and the natural gas price was $4/MMBtu. Annual simulation of the fuel cell system without any maintenance shut down shows the annual fuel cell system cogeneration efficiency (LHV basis) to be 84.3%. Over the course of the year, the SOFC met 91% of the total house electric energy requirement. On the thermal side, the fuel cell system was able to provide for 54% of the total annual domestic hot water energy requirements.

Table 3 summarizes the economic performance of the 2 kWe SOFC residential power system against the base case of utility provided electricity

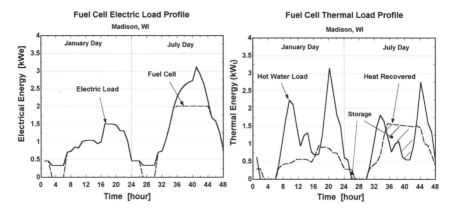

Figure 12. Fuel Cell Operating Profiles.

and gas. From a simple energy usage viewpoint (i.e., no fixed transmission and distribution costs), the use of the residential SOFC resulted in electric utility savings of $563, a 90% reduction. However, in the case where no heat was recuperated, the gas utility requirement increased 169% from $123 to $331 due to SOFC fuel consumption. A simple payback of nearly 11 years would result from an electric only operation. With cogeneration in the form of domestic hot water, the utility savings could be increased by nearly 20% over the electric-only system achieving a 9.8 year payback.

The *electric* capacity factor* of the fuel cell is an important performance parameter as it measures the total annual operating usage of the high capital cost component. The larger the capacity factor, the better the fuel cell payback economics appear. The electric capacity factor performance of the

Table 3. Economic Summary of 2 kW SOFC Residential Cogenerator

System	Electric ($)	Gas ($)	Total Cost ($)	Payback (yrs)
Grid electricity; gas-fired water heater	622	123	745	(base case)
SOFC + grid backup; gas-fired water heater	59	331	390	10.9
SOFC + grid backup; cogeneration with gas-assist	59	265	324	9.8

* The annual fuel cell electric capacity factor is defined as the kWh supplied by the fuel cell divided by the maximum kWh it could have supplied (e.g, 2 kW * 8,760 hours.)

fuel cell indicates that a 2 kW size solid oxide fuel cell may be too large as only 46% of its annual electrical energy production capacity was utilized. Employing a smaller fuel cell system of 1 kW could conceivably double the electric capacity factor to 92%. Other methods to increase the fuel cell electric capacity factor include the use of lead acid batteries, heat pumping, and where possible, selling electricity back to the grid for "net metering."

The *thermal* capacity factor* of the fuel cell is also a useful measure. An 81% thermal capacity factor was achieved in cogeneration mode displacing 4800 kWh of thermal energy that otherwise would have been served by the hot water heater. This high degree of waste heat recovery was possible due to the use of a 2–tank thermal storage configuration. Use of both electrical and thermal storage would effectively increase fuel cell electric and thermal capacity factors. Additionally, this may enable more high efficiency baseload operation strategies for the SOFC. In fact, such high efficiency operation may be necessary to offset the inefficiencies of power conditioning and electrical storage requirements of battery-included systems.

CONCLUSIONS

A method for optimal fuel cell operating point selection for system design was presented in the first part. The results indicated the optimal operating cell voltage to be 700 mV and an optimal operating temperature of 800°C. This analysis showed an optimal fuel utilization of 89%, but further study of fuel utilization may be required. The use of external reforming in solid oxide fuel cell systems is one design option that is likely to take place in some near term commercial units [27]. Such systems require additional cooling air and may be more capital cost intensive. System optimization and performance comparisons between internal and external reforming-based systems need to be made. Finally, the operating point analyses presented herein were based on electric-only systems. Inclusion of a heat credit for cogeneration fuel cell power plants is another area for further study in selection of operating points.

A simulation of a 2 kW residential SOFC power generator was carried out. Results indicated an annual cogeneration efficiency of 84% and a ten-year simple payback was possible. The meeting of electric and hot water loads by the fuel cell system was accommodated by a 2–tank hot water storage system. The low electric capacity factor of 46% indicates that a 2 kW rating for the home in this analysis may be too large. Electrical storage,

---

* The thermal fuel cell capacity factor is defined as the kWh recovered from the exhaust gas divided by the kWh that could have been supplied had the exhaust gases been reduced to the water main temperature.

heat pumping, and net metering are a conceivable options to boost electric capacity factor, however, optimal operating and control methodologies must be devised to maximize efficiency and economic performance.

ACKNOWLEDGEMENTS

The author would like to thank Professors Sandy Klein and Doug Reindl for their guidance, unswerving support, and many helpful discussions. I would also like to thank Professors William Beckman and John Mitchell for their useful suggestions. Finally, I must express my sincere appreciation to the Link Foundation and the Energy Center of Wisconsin for providing financial support.

REFERENCES

[1] R.J. Braun, S.A. Klein, D.T. Reindl, "Review of State-of-the-Art Fuel Cell Technologies for Distributed Generation," Report 193–2 prepared for the Energy Center of Wisconsin, January (2000).

[2] P. Schafer, "Commercial Sector Solid Oxide Fuel Cell Business Assessment," Electric Power Research Institute, TR-106645, August (1996).

[3] T. Chen, J.D Wright, K. Krist, "SOFC System Analysis," *Proc. Of Fifth International Symposium on Solid Oxide Fuel Cells* (SOFC-V), PV97–18, Germany (1997).

[4] E. Riensche, U. Stimming, G. Unverzagt, "Optimization of a 200 kW SOFC Cogeneration power plant, Part I: Variation of process parameters," *J. Power Sources*, **73** (2), pp. 251–256 (1998).

[5] A. Khandkar, J. Hartvigsen, S. Elangovan, "A Techno-Economic Model for SOFC Power Systems," to be published in *Solid State Ionics.*

[6] R.J. Braun, S.A. Klein, and D.T. Reindl, "Operating Point Analyses of SOFC Energy Systems," *Proc. of the 4th European SOFC Forum*, Lucerne, Switzerland, pp. 459–468 (2000).

[7] W.L. Lundberg, "Solid Oxide Fuel Cell Cogeneration System Conceptual Design," Final Report, GRI-89/0162, Gas Research Institute, Chicago, IL, July (1989).

[8] J.R. Rostrup-Nielsen, "Catalytic Steam Reforming," in *Cat. Sci. and Tech.*, **5**, Springer-Verlag, New York, p. 83 (1984).

[9] E.S. Wagner and G.F. Froment, "Steam Reforming Analyzed," *Hydrocarbon Processing*, July, pp. 69–77 (1992).

[10] U.G. Bossel, "Solid Oxide Fuel Cells Data Book, Facts & Figures," International Energy Agency SOFC Task Report, Swiss Federal Office of Energy, Berne (1992).

[11] J.H. Hirschenhofer, D.B. Stauffer, and R.R. Engleman, "Fuel Cells, A Handbook," 4th Edition, prepared for the U.S. Department of Energy, November (1998).

[12] J.J. Hartvigsen and A.C. Khandkar, "Thermally Integrated Reformer for Solid Oxide Fuel Cells," U.S. Patent No. 5,366,819, Washington, D.C., November (1994).

[13] S.C. Singhal, "Recent Progress in Tubular SOFC Technology," *Proc. of the Fifth International Symposium on Solid Oxide Fuel Cells* (SOFC-V), PV97–18, Germany, The Electrochemical Society (1997).

[14] Engineering Equation Software (EES), F-Chart Software, Middleton, WI.

[15] A.J. Appleby, "Characteristics of Fuel Cell Systems," in *Fuel Cell Systems*, Edited by L.J. Blomen and M.N. Mugerwa, Plenum Press, New York (1993).

[16] D. Ghosh, G. Wang, R. Brule, E. Tang, and P. Huang, "Performance of Anode Supported SOFC Cells," *Proc. Of the Sixth International Symposium on Solid Oxide Fuel Cells* (SOFC-VI), Honolulu, HI, The Electrochemical Society (1999).

[17] N. Minh, "Development of Thin-film SOFCs for Power Generation Applications," *1994 Fuel Cell Seminar, Program and Abstracts*, San Diego, CA, pp. 577–580 (1994).

[18] Michael Hsu, "Advanced Planar SOFC Development," Electric Power Research Institute TR-107116, Oct. (1996).

[19] R.F. Boehm, *Design Analysis of Thermal Systems*, John Wiley & Sons, New York (1987).

[20] *Perry's Chemical Engineer's Handbook*, 7th Edition, McGraw-Hill, Inc., New York (1997).

[21] M.S. Peters and K.D. Timmerhaus, *Plant Design and Economics for Chemical Engineers*, 3rd Edition, McGraw-Hill, Inc., New York (1980).

[22] S. Linderoth and M. Mogensen, "Improving Durability of SOFC Stacks," *Proc. Of the 4th European SOFC Forum*, Lucerne, Switzerland, pp. 141–150 (2000).

[23] K. Krist and J. Wright, "SOFC Residential Cogeneration," *1999 Joint DOE/EPRI/ GRI Fuel Cell Technology Review Conference*, Chicago, IL, August (1999).

[24] J.J. Mutch "Residential Water Heating, Fuel Consumption, Economics, and Public Policy," *RAND Report R1498* (1974).

[25] S.A. Klein, W.A. Beckman, and J.A. Duffie, *TRNSYS—A Transient System Simulation Program*, Solar Energy Laboratory, Madison, WI (1999).

[26] Peter Bos, "Commercializing fuel cells: managing risks," *J. Power Sources*, 61, 21–31 (1996).

[27] A. Khandkar, S. Elangovan, J. Hartvigsen, D. Rowley, and M. Tharp, "Recent Progress in SOFCO's Planar SOFC Development," *1998 Fuel Cell Seminar Abstracts*, Palm Springs, CA, November, pp.465–468 (1998).

[28] R.A. Gaggioli, S.D. Moody, and W.R. Dunbar, "Optimization of SOFC Processes by Exergy Analysis," *Proc. of the 1st European SOFC Forum*, Lucerne, Switzerland, pp. 129–152 (1994).

Lithium Salts of Carboranyl Anions $CB_{11}R_{12}^{-}$ as Novel Components for Solid Polymer Electrolytes

Ilya Zharov

Department of Chemistry and Biochemistry

University of Colorado at Boulder,

Boulder, CO 80309

Research Advisor: Dr. Josef Michl

ABSTRACT

An approach to a new family of lithium ion conducting polymers based on the lithium salts of alkylated carboranyl CB_{11} anions is described. A number of substituted $CB_{11}R_{12}^-$ anions suitable for the incorporation into polymers has been prepared, as well as several conducting materials based on these anions. Preliminary results show that such materials can be made relatively easily and that they promise to provide superior solid polymer electrolytes.

INTRODUCTION

Lithium rechargeable batteries based on solid polymer electrolyte (SPE) technologies are presently proposed for a wide variety of such demanding applications as electric vehicle, start-light ignition, and portable electronic and personal communication [1]. These applications have very specific and differing sets of performance requirements. For example, electric vehicle applications are likely to require high energy density, high pulse power and extremely long cycle life at deep depth of discharge. On the other hand, start-light ignition battery applications require high pulse power over a wide range of temperature and long cycle life at shallow depth of discharge. Finally, portable electronics and personal communication applications require highly reliable and safe batteries with high energy density, high specific energy and long cycle life.

Liquid electrolyte batteries presently used have a number of serious disadvantages [2]. The liquid electrolytes are not entirely stable chemically (e.g, react with lithium anodes and cathode materials) nor electrochemically (i.e., can undergo electrolysis). In addition, the always possible leakage makes liquid electrolyte batteries both unreliable and environmentally unsafe. Finally, their processability is limited, since very desirable thin film production and/or a variety of shapes can not be achieved for such batteries. In this respect the lithium polymer electrolyte technology could provide a highly flexible and versatile approach, although it is unlikely that a single lithium polymer electrolyte rechargeable battery could meet all of the specific requirements mentioned above. However, there are performance parameters which are common to all lithium polymer electrolytes and which will be required to ensure technological success in any application [3]. These include a high ionic conductivity at room temperature, high transport number for the cation and an amorphous character of polymers, good mechanical and dimensional stability, and easy processability for thin film production.

Thus far lithium polymer electrolyte rechargeable batteries have exhibited several important performance limitations [4]. They have not met successfully all of the power density, energy density, specific energy, low temperature performance, and cycle life expectations. The current approaches provide little advantage over liquid electrolyte systems. There is no polymer electrolyte system that simultaneously provides ionic conductivity higher than 10^{-3} S/cm at ambient temperature, lithium cationic transference number above 0.9, and good mechanical properties. There is no polymer electrolyte which has demonstrated chemical and electrochemical stability in contact with lithium anodes and/or high voltage cathode materials. Processing and manufacturability of polymer electrolyte components remain to be demonstrated on a commercial scale.

Methods of maximizing conductivity and other properties of SPE's include modifications of both the polymer host and the salt. A large number of polymers have been described and characterized. However, it is possible to group all of the polymer systems within only two classes: (i) pure solid polymer electrolyte systems, and (ii) network or gel-polymer electrolyte systems [5]. Evolution of polymer systems for SPE can be described in terms of several generations [6].

Following the discovery by Wright [7] in 1973 that complexes formed between polyethylene oxide (PEO) and various alkali metal salts exhibited high ionic conductivities at elevated temperatures, and Armand's proposal for an all-solid-state lithium battery [8], a large number of groups began to study PEO and its modifications. These systems have such advantages as good mechanical and electrical properties, large redox stability windows, good compatibility with cathodes and lithium anode, a very high solvating power and chain flexibility at elevated temperatures [9]. However, it was found that a number of serious setbacks exist for PEO-salt systems, such as low conductivity at ambient temperature, very low cation transference numbers and high crystallinity [10].

Several approaches were undertaken via modification of the polymer chain to increase the amorphous character of PEO-type polymers in an effort to lower the T_g and increase the ionic conductivity and transport number of the cation at room temperature. Examples of such polymers are random copolymers [11], block copolymers [12], comb-branched block copolymers [13], networks [14], and single ion conductors [15]. These systems include copolymers made of polyether grafted polyethers, polysiloxanes and polyphosphazenes, which provide good mechanical properties and good chemical stability, but still suffer from a relatively low conductivity.

The gel-polymer electrolytes [16] are characterized by a higher ambient ionic conductivity but poorer mechanical properties when compared with the above pure polymer electrolyte systems. The gel electrolytes are obtained by addition of a large quantity of liquid plasticizer and/or solvent to a polymer matrix capable of forming a stable gel with a polymer host structure. To improve mechanical properties of the gel-polymer electrolytes, components which can be cross-linked are added to the formulation. Consequently, the overall ionic conductivity of gel-polymer electrolytes is much closer to ionic conductivity in liquid aprotic solvents, but their electrochemical stability is also similar to that of liquid based electrolytes.

More recently quasi-solid state concepts were proposed by Angell [17], Peled [18], Ogata [19] and Scrosati [20]. For example, the work of Angell involves the use of molten conducting salts with the polymer as a minor component, imparting mechanical integrity to thus formed "ionic rubbers". In composite electrolytes [21] ionic conductivity is enhanced through the addition of an insoluble second phase, such as aluminum oxide or silica, to crystalline electrolytes.

As can be seen from the above brief historical overview, most emphasis in solid polymer electrolyte development was made on modifications in polymers and additives, and a relatively limited number of salts were investigated. Yet, the nature and concentration of the incorporated salt may have a major influence on the properties of SPE. Essentially, polymer electrolytes form when the salt consists of a polarizing cation and a large anion of delocalized charge to minimize the lattice energy. The salt affects the conductivity through crystalline complex formation, intramolecular cross-linking of the polymer chains, and the degree of salt dissociation (the number of charge carriers). In particular the anion has a major influence on phase composition and conductivity, as has been shown for such commonly used anions as perchlorate and triflate, and others [22].

It occurred to us that using highly alkylated or peralkylated carboranyl anions [23] may improve a number of major characteristics of SPE materials, such as ionic conductivity, chemical and electrochemical stability, processability and safety, while the preparation of SPE materials containing those anions may be quite easy. Structurally, the anions we wanted to work with are derivatives of the long-known icosahedral monocarba-*closo*-dodecaborate anion $CB_{11}H_{12}^-$ (**1**) [24]. Our assumptions were based on a number of unique properties of $CB_{11}Me_{12}^-$ (**2**) whose structure, thermal behavior, solubility [25], solution conductivity, and chemical and electrochemical stability [26] have been studied.

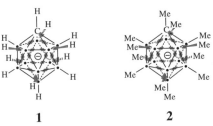

1   2

The mobility of lithium cations in polymer electrolytes is a crucial factor enabling fast and/or deep discharge of polymer electrolyte based lithium rechargeable batteries [3]. Low cationic transference numbers, especially in conjunction with low electrolyte conductivity, can adversely affect cell performance. One approach to improve lithium cationic transference number in polymer is to develop large low-basicity anions with delocalized charge to minimize ion pairing and lattice energy [22]. So far, ClO_4^-, $CF_3SO_3^-$ and PF_6^- are the only common anions that fulfill these requirements to some extent. It has been shown [27] that $CB_{11}Me_{12}^-$ (**2**) is one of the least nucleophilic anions known today [28]. It is quite large (the van der Waals radius of the anion **2** is ca. 10 Å) and its negative charge is delocalized inside the carboranyl cage with no exposed electron pairs [25]. Such an anion is likely to move very slowly in the polymer solutions, increasing cation transference numbers by decreasing concentration gradients.

Increasing the salt concentration is yet another way to improve the ion conductivity of solid polymer electrolytes. Often this is limited by a salt solubility in polymers and/or additives. It is especially important in the case of polysiloxanes, often superior polymers for SPE applications due to their low T_g and a low oxygen bacisity, but also usually poor solvents for alkali metal salts. We found that the lithium salt of **2** is soluble in polymers such as poly(dimethylsiloxane) or poly(methyphenylsiloxane) up to 20% [29]. Even prior to any optimization the conductivity of the latter solutions is 5×10^{-5} S/cm at 25 °C [29] which is comparable to the SPE materials produced so far [30].

Another approach to improve lithium cationic transference number in polymer is to develop a highly conductive, cation-conductive ionomer [5]. Such polyelectrolyte would ensure unity transference number for the cation. However, results of previous investigations indicate that ion pairing limits the performance of these materials [22]. Incorporation of $CB_{11}Me_{12}^-$ into polymeric structures could greatly improve conductivities of the above polyelectrolyte due to the very low nucleophilicity of this anion and thus a significantly weaker anion-cation interaction.

In polymer-salt solutions the stoichiometric crystalline phase and a more dilute amorphous phase are present in an equilibrium above T_g, and the amorphous phase is responsible for the ionic conductivity [31]. The presence of lithium salts with low lattice energy (associated with a large anion that would effectively delocalize the negative charge) allows the polymer to be composed exclusively of amorphous phases [32]. Symmetry of the anion or flexibility of its chain can introduce an additional plasticizing effect that lowers T_g and thus enhances conductivity [33]. In this respect the highly symmetrical large $CB_{11}R_{12}^-$ anions looked very promising.

Chemical and electrochemical stability are two other very important factors for SPE materials. The dry lithium salt of $CB_{11}Me_{12}^-$ is thermally stable in air at least to 300°C. It is stable toward strong bases and does not react with lithium metal. The lithium salt of **2** could also afford a very high electrochemical stability. Indeed, in acetonitrile, the anion $CB_{11}Me_{12}^-$ (**2**) undergoes a reversible one-electron oxidation at 1.6 V against an Ag/AgCl electrode and yields the stable neutral radical, $CB_{11}Me_{12}^\bullet$ [27]. In the polymer and against a lithium electrode this value is expected to be to over 4.6 V.

We expected the synthesis of the new structures to be very simple, at least judging by the facile access to the first member of the family $CB_{11}Me_{12}^-$

(2), and using the information about alkylations and other substitutions of 1 accumulated in the Michl research group.

It is known that C-substitution on 1 can be achieved by lithiation of position 1 with butyllithium and reaction with a suitable reagent [34]. This operation has to be performed before hydrogens in positions 2–12 are replaced by methyls (4) since attempts to replace hydrogen in position 1 of the 2–12 undecamethylated anion with lithium by action of n-BuLi or t-BuLi failed, presumably as a result of a combination of steric hindrance with reduced acidity due to electron donation by the eleven methyl groups.

Position 12 in 3, opposite to the carbon atom, can be iodinated (5) [35]. Iodine can be replaced with an alkyl group, such as Et, Bu, i-Pr, in a reaction with an alkyl Grignard reagent catalyzed by palladium (6) [36]. Compound 5 can be alkylated with methyl triflate (7) and iodine removed by reduction, yielding the Me_{11}–anion 8 [37]. Also, all of the positions 7–12 of 3 can be iodinated (9) following a literature procedure [38]. Subsequent methylation in positions 2–6 (10) and reductive removal of iodines lead to the Me_6–anion 11 [37].

A maximum of nine ethyl groups can be introduced into 1 using ethyl triflate [37]. Six additional ethyl groups can be introduced into positions 7–12 of 1–ethyl substituted 1 (12) [37]. The methyl group in position 12 of the permethylated anion 2 can be replaced by a fluorine atom by reacting 2 with anhydrous HF [39]. Recent efforts in our group on substitution in carboranyl anions provided further useful information. For example, the number of useful C-substituted anions has been extended considerably (13). They now include those with a phenyl group, prepared by Negishi coupling, with trimethylstannyl group, and with a boronate residue, prepared in analogy to the reported C-alkylation, and with chlorine, bromine and iodine, prepared by the reaction of C-lithiated anion either directly with the halogen (in the case of iodine) or via the corresponding copper reagent (in the case of chlorine and bromine). It was discovered that heating the lithium salt of 2 with anisole or benzene introduces a phenoxy group (14) or a phenyl group (15) into position 12, respectively. While the derivatives of 1 halogenated in position 1 cannot be permethylated, the boronic ester in position 1 can (16). Also, six methyl groups can be selectively introduced into positions 7–12 of 1 (17) by first introducing a triisopropylsilyl group into position 1 of 1 and subsequently methylating $1-i-Pr_3Si-CB_{11}H_{11}^-$ and removing the protecting group with CsF in sulfolane. We also found that 2 can be converted into the 12–hydroxy derivative 18 by heating the H_3O^+ salt of 2 under vacuum. The perfluorination of 2 has recently been accomplished [40]. Finally, work is under way in the Michl research group to improve the synthetic starting point.

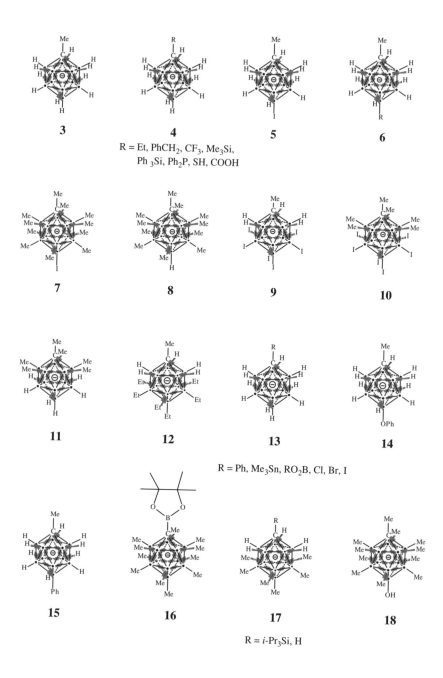

R = Et, PhCH$_2$, CF$_3$, Me$_3$Si,
Ph$_3$Si, Ph$_2$P, SH, COOH

R = Ph, Me$_3$Sn, RO$_2$B, Cl, Br, I

R = i-Pr$_3$Si, H

OBJECTIVE

The objective of this work was to prepare and evaluate new SPE materials based on covalently attached $CB_{11}R_{12}^-$ anions. We planned to start with two simple approaches to the preparation of these materials.

1. To produce polymeric materials with the anions in side chains one could prepare monomers by introducing polymerizable substituents into the anion, such as an alkenyl group, for example on the carbon atom of 2 (e. g., 19) or on the oxygen atom of 18 (e. g., 20). Their polymerization and/or copolymerization (e. g., with styrene) could be achieved by the use of transition metal catalysts (e. g., 21) [41]. One could also introduce dialkoxysilyl substituents, which can be easily polymerized to silicones by hydrolysis (e. g., 22) [42].

2. To incorporate $CB_{11}R_{12}^-$ anions directly into existing polymers one could react the lithium salt of 2 with with polymers containing phenyl groups, such as polystyrene or poly(phenylmethylsiloxane). This could yield a polymer with $CB_{11}Me_{11}^-$ units attached to the side chains via phenyl links (23 and 24). One could also react the hydroxy anion 18 with functionalized polymers such as Merrifield resin to produce a polymer with the anions incorporated into side chains via oxygen links (25).

RESULTS

Preparation of $CB_{11}R_{12}^-$ Anions and of the Monomers

The original permethylation procedure utilized methyl triflate and the quite expensive di(tert-butyl)pyridine as a base [25]. An improved procedure uses calcium hydride and sulfolane at room temperature [43]. Its disadvantage is that it takes 18–25 days to complete. Our attempts to accelerate this reaction by raising the temperature resulted in undesirable triflate substitution of the carboranyl cage. Now we found that decreasing the amount of sulfolane significantly speeds up the reaction. When two moles of sulfolane are used per one mole of methyl triflate, the permethylation reaction of 3 is complete in five days at room temperature. In addition, other alkylations could now also be performed much better. For example, a small scale nonaethylation of $CB_{11}H_{12}^-$ was complete in only about one week, while before this reaction required 4.5 months [43].

Monomers 19 and 20 were chosen to be prepared and examined first, based on the expected relative ease of synthesis. Additionally, it could be possible to convert the monomers 19 and 20 into a trialkoxysilyl derivatives which could be used to prepare the corresponding polysiloxane. Monomer 20 could be modified to provide other polymers.

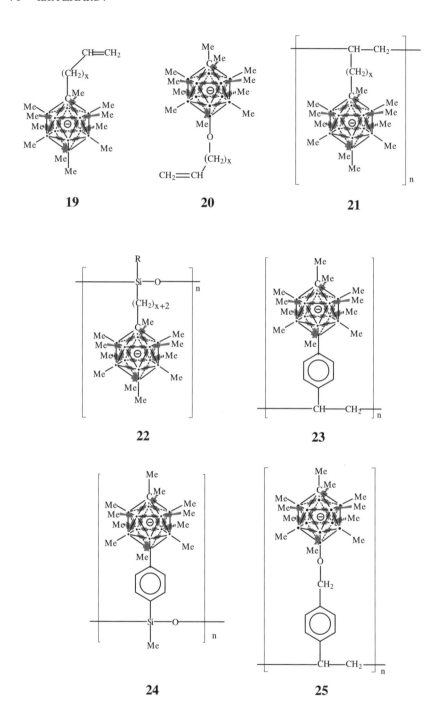

19

20

21

22

23

24

25

Our first attempt to prepare the monomer **19** is shown in Scheme 1. The first step to produce **26** has been previously described [43]. Unfortunately, in the second step the double bond of **26** was also methylated.

In order to avoid this complication, a 6–chlorohexyl group was introduced into position 1 of the carboranyl cage (**27**, Scheme 2). This reaction proceeds in a good yield. Compound **27** was then successfully permethylated to yield the alkene precursor **28** (Scheme 2).

The elimination required for the next transformation is well precedented. We tested many known elimination conditions, utilizing reagents such as EtOK in EtOH, potassium ethylene glycolate in ethylene glycol, *t*-BuOK in *t*-BuOH, and CsF in DMF. All of these reactions suffered from the formation of side products, mainly as a result of substitution. Finally, it was found* that a Schwesinger base provides the desired monomer **19** cleanly in the case of 6–bromohexyl analog of **28** (**29**, Scheme 3).

We found that the preparation of the permethylated hydroxy anion **18** is possible by heating the H_3O^+ salt of **2** under vacuum. Unfortunately, this reaction provides **18** in a mixture with the starting material, and their separation proved to be a formidable task. Later, we found that 1–Me-$CB_{11}H_{11}^-$ (**3**) reacts with neat dry triflic acid to provide 1–Me-12–OTf-$CB_{11}H_{10}^-$. This compound can be permethylated at elevated temperatures to provide 12–OTf-$CB_{11}Me_{11}^-$, which can be reduced to produce the desired **18** (Scheme 4). Finally, we were able to further improve the preparation of **18** by finding that 1–Me-$CB_{11}H_{11}^-$ (**3**) reacts with methyl triflate at room temperature in the absence of sulfolane to give 1–Me-12–OTf-$CB_{11}H_{10}^-$, which can be then permethylated at high temperature, thus providing 12–OTf-$CB_{11}Me_{11}^-$ in a single-pot process (Scheme 4).

We studied the reaction of **18** with benzyl chloride as a model for its incorporation into poly(*p*-chloromethylstyrene) and found (by ES/MS) that in THF with NaH and NaI this reaction goes to completion (Scheme 5). Unfortunately, the analogous reaction with vinylbenzyl chloride did not produce the desired product **30** under the conditions we tried (e. g., NaH/NaI/THF and *n*-BuLi/THF) (Scheme 5).

We also tested a number of coupling conditions (e. g., NaH/NaI/THF, n-BuLi/THF) with 6–bromopentene but were not able to prepare the monomer **20**. At the same time, we found that methyl and ethyl triflates react with **18** to give the corresponding alkoxy derivatives (Scheme 6) and subsequently demonstrated that pentenyl triflate converts **18** to **20** (Scheme 6).

---

*Work conducted by Dr. Photon Rao.

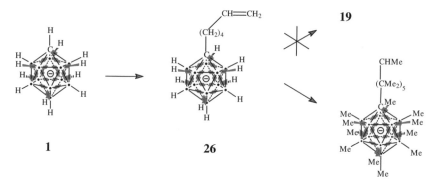

Scheme 1

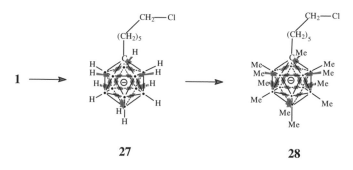

Scheme 2

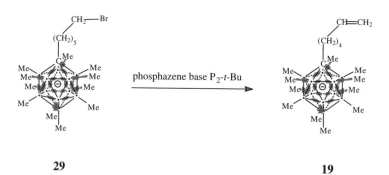

Scheme 3

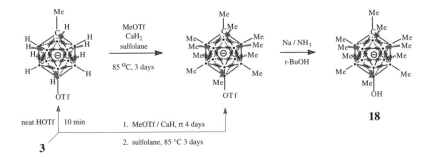

neat HOTf | 10 min   1. MeOTf / CaH, rt 4 days

                          2. sulfolane, 85 °C 3 days

3

18

Scheme 4

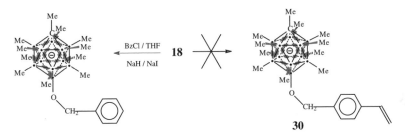

Scheme 5

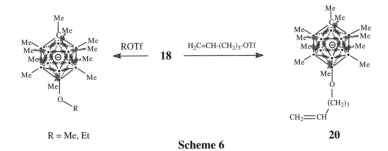

R = Me, Et

Scheme 6

Preparation of the Polymers

Monomer **19** was polymerized by Dr. Photon Rao using a zirconium catalyst,[44] [Cp*$_2$ZrMe]$^+$ [B(C$_6$F$_5$)$_3$Me]$^-$, to provide a mixture of low molecular weight oligomers **21** (Scheme 7).

The olefin **26** was converted into the corresponding trialkoxysilyl derivative, which was successfully polymerized (Scheme 8) to yield the gel **31**. This reaction sequence was carried out by collaborators in France (laboratory of Prof. R. Corriu in Montpellier), to whom we sent a sample of **26**. However, so far our French collaborators were not able to convert **27** into the alkoxysilyl substited monomer.

We also tried a direct incorporation of **18** into a functionalized polymer. We used poly(vinylbenzyl chloride) and a Merrifield resin under the conditions described above, but were not able to isolate the desired polymeric product **25** (Scheme 9).

It was mentioned earlier that the lithium salt of **2** reacts with benzene resulting in a replacement of one of the methyl groups by a phenyl group. We utilized this reaction in order to produce polymers **23** and **24** by heating Li$^+$ **2** with the corresponding polymers, i. e., polystyrene and poly(phenylmethylsiloxane).

Conductivity of the Produced Materials

The following samples have been given to Eagle-Picher, LLC for AC impedance conductivity measurements:

1. A mixture of low molecular weight oligomers **21** produced from **19**.

2. A solution of Li$^+$ **2** in poly(phenylmethylsiloxane) prepared by dissolving the lithium salt (dried under vacuum at 150°C) in the polymer.

3. Poly(phenylmethylsiloxane) doped with Li$^+$ **2** prepared by heating the solution of the Li+ salt in the polymer under vacuum for 3 days at 160°C.

4. Polystyrene doped with Li$^+$ **2** prepared in the same manner as (3).
   Sample no. 1 showed a conductivity of 2.21 × 10^{-3} S/cm.
   Sample no. 2 showed a conductivity of 1.41 × 10^{-4} S/cm.
   Sample no. 3 showed a conductivity of 1.62 × 10^{-4} S/cm.

The numbers obtained so far are very promising. Sample no. 4 is being presently tested. In addition, work is under way on measuring the conductivity of the solid gel **31**.

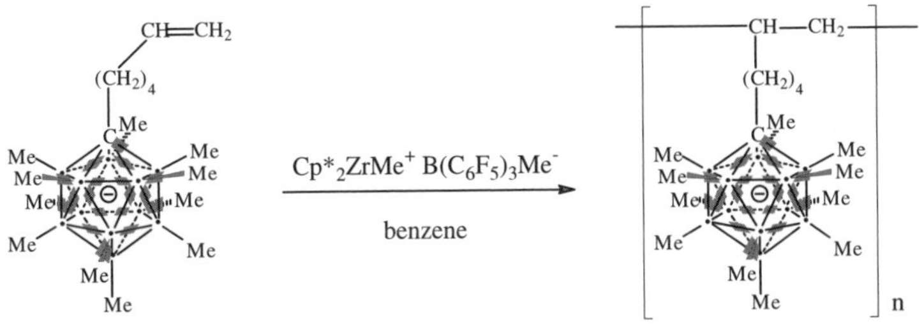

19   **21**

Scheme 7

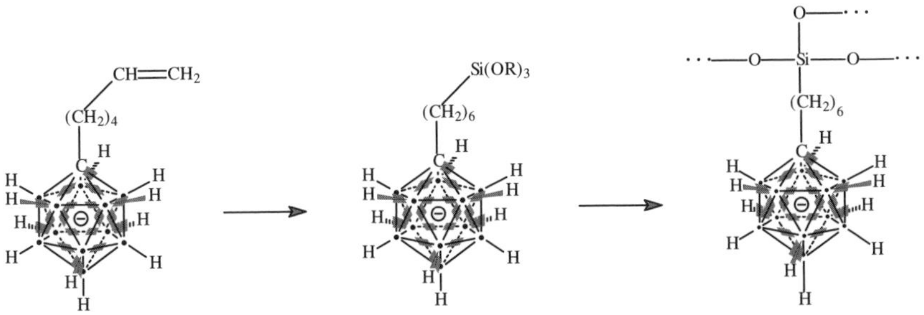

26   **31**

Scheme 8

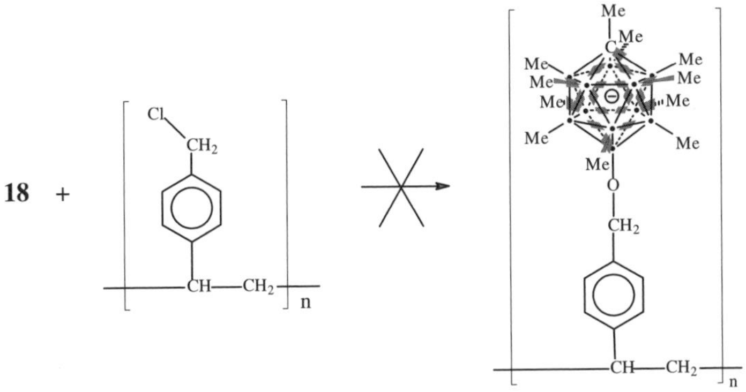

18 +

25

Scheme 9

CONCLUSIONS AND FUTURE WORK

We described an approach to a new family of lithium conducting polymers based on the lithium salts of alkylated carboranyl CB_{11} anions. Our preliminary results showed that such materials can be made relatively easily and that they promise to provide solid polymer electrolytes with superior properties.
Future work could take a number of directions suggested below:

1. The preparation of the monomer **20** could be further optimized.
2. Monomers other than **19** and **20** could be explored. They could include (a) anions with longer alkyl chains, (b) anions with position 12 protected by either a bulky alkyl group or a fluorine atom, (c) anions containing trifluoromethyl groups, (d) anions with aryl substituents. In addition, work should be continued on the preparation of the monomer **30**, which we unsuccessfully attempted to synthesize from the hydroxy anion **18**.
3. Conditions of the transition metal catalyzed polymerization we used so far could be optimized. For example, a new catalyst,[45] based on tridentate bis-imine ligands coordinated to iron and cobalt, should be tested. Also, new types of polymerization, including alkene metathesis, anionic and radical polymerization, could be used.
4. The work started on the direct incorporation of carboranyl anions into functionalized polymers could be continued.
5. The properties of the resulting polymeric materials should be studied, including their thermal properties (T_g), viscosity, and molecular weight distribution. Those could be adjusted by copolymerizing our monomers with standard monomers such as styrene. The composition of the polymeric materials should be adjusted based on the AC impedance measurement results. For example, we know that addition of a small amount of water increases the conductivity of the benzene solution of Li^+ **2** by two orders of magnitude. Water is incompatible with lithium cathodes, but the result suggests that polar functional groups, such as nitrile, should perhaps be introduced into the polymers.
6. New types of polymers containing carborane units in the backbone could be examined.

EXPERIMENTAL

General

$Me_3NH^+ CB_{11}H_{12}^-$ was purchased from Katchem Ltd., E. Krásnohorské 6, 11000 Prague 1, Czech Republic. Solution ^1H, ^{11}B, and ^{13}C spectra were measured with a Varian XRS-300 and a Varian Unity-500 spectrometers. ^1H and ^{13}C chemical shifts were measured relative to the lock solvent. ^{11}B chemical shifts were measured relative to BF_3OEt_2, with positive chemical shifts downfield. $B(OCH_3)_3$ was used as an external standard (18.1 ppm). Mass spectra were recorded with a Hewlett Packard 5989 API/ES/MS. An HPLC system employing a reverse phase C_{18} column (250 × 4.6 mm, 5 μ) with methanol/water (containing a 1% AcOH / 0.7% Et_3N buffer) as the mobile phase was used for monitoring reactions, while larger columns of the same phase were used for semi-preparative separations. All volatile compounds were analyzed using Varian 3400 analytical GC with 0.2 mm × 25 m 0.33 μm RSL-150 5% crosslinked silica capillary column. Their separations were performed using Varian 3400 preparative GC with 1/4" × 21' 5% OV-7 80/100 Chromosorb GHP packed column. Individual compounds were purified to >99.5% purity (by analytical GC).

Improved Permethylation Procedure

Calcium hydride (8.4 g, 200 mmol) is added to a solution of Cs^+1–Me-$CB_{11}H_{11}^-$ (1.45 g, 5 mmol) in 19 mL of sulfolane (24 g, 200 mmol, freshly distilled *in vacuo* from CaH_2) in a 250–mL, three-necked, round-bottom flask, and the system is placed under argon. The mixture is stirred at 25°C and methyl triflate (11.3 mL, 16.4 g, 100 mmol) is added to the flask with a syringe pump over a period of 20 h. The stirring is continued for 3 days, after which the reaction mixture solidifies. A solution of 5.5 mL of methyl triflate in 10 mL of sulfolane is added to the reaction mixture in one portion and stirring is continued for 2 days, after which the reaction is complete. The solidified reaction mixture is diluted with 300 mL of dry methylene chloride and CaH_2 is removed by vacuum filtration. The filtrate is quenched slowly with 100 mL of 27% ammonium hydroxide solution and is extracted three times with 150 mL of diethyl ether. Calcium hydride is carefully dissolved in 15% HCl (100 mL), the solution is neutralized with ammonium hydroxide and extracted three times with 50 mL of diethyl ether. The combined organic layer is washed with 20% aqueous CsCl (3 × 50 mL). The combined CsCl wash is extracted with diethyl ether (3 × 50 mL). The combined ether solution is evaporated. The residual sulfolane is removed by a short path vacuum (<0.05 mm Hg) distillation at 150°C. The white solid collected is $Cs^+CB_{11}Me_{12}^-$ (2.0 g, 90% yield, 99% pure by HPLC).

A solution of $Me_3NH^+CB_{11}H_{12}^-$ (20.3 mg, 0.1 mmol) in sulfolane (0.36 g, 3 mmol) was stirred with EtOTf (0.53 g, 3 mmol) in the presence of CaH_2 (0.13 g, 3 mmol) at r.t.. The reaction was monitored by ES/MS. After 1 week a mixture containing mainly a product with m/e 395 ($CB_{11}Et_9H_3^-$) was observed.

Cs⁺ 12-HO-CB₁₁Me₁₁⁻ (18)

Cs⁺ 12-HO-CB$_{11}$Me$_{11}^-$ (18)

Calcium hydride (8.4 g, 200 mmol) is combined with methyl triflate (11.3 mL, 16.4 g, 100 mmol) in a 250–mL, three-necked, round-bottom flask, and the system is placed under argon. Cs⁺1–Me-CB$_{11}$H$_{11}^-$ (1.45 g, 5 mmol) is added and the mixture is stirred at 25°C for 4 days, after which it solidifies. Sulfolane (30 mL 36.0 g, 300 mmol, freshly distilled *in vacuo* from CaH₂) is added to the reaction mixture and stirring is continued for 3 days at 90°C, after which the reaction is complete. The resulting Cs⁺12–TfO-CB$_{11}$Me$_{11}^-$ is isolated as described above. This compound (1.16 g, 2 mmol) is added to a solution of 0.09 g (4 mmol) of sodium in 50 mL of liquid ammonia at –39°C. The reaction ixture is stirred at that temperature for 40 min, after which *tert*-butanol (0.5 mL) is added. The reaction mixture is allowed to warm up to room temperature and ammonia is allowed to evaporate. The solid residue is dissolved in water and the solution is extracted with diethyl ether (3 × 150 mL). The organic layer is washed with 20% aqueous CsCl (3 × 50 mL). The combined CsCl wash is extracted with diethyl ether (3 × 50 mL). The combined ether solution is evaporated. The resulting white solid is crystallized from water to give 0.82 g (92%) of pure **18**. ¹H NMR (acetone-d_6): δ 0.817 (3 H, s, 1–CH₃), –0.344 (30 H, s, 2–11–CH₃). ¹¹B {¹H} NMR (acetone-d_6): δ 9.32 (B₁₂), –10.15 (B$_{7-11}$), –12.80 (B$_{2-6}$). ¹³C {¹H, ¹¹B} NMR (acetone-d_6): δ 49.02 (1–C), 12.92 (1–CH₃), –3.71 (7–11–CH₃), –4.07 (2–6–CH₃). IR (KBr pellet): 909, 1043, 1153, 1312, 1428, 2827, 2854, 2891, 2928, 3245 cm⁻¹. MS/ES(-) in methanol: base peak at m/e 313 with the expected isotopic distribution (M⁻). Anal. Calcd for C₁₂H₃₄B₁₁CsO: C, 32.30; H, 7.68. Found: C, 32.16; H, 7.83.

Compound **18** was also produced when the H₃O⁺ salt of CB₁₁Me₁₂⁻ (**2**), prepared by adding HCl/ether solution to Ag⁺ **2** in ether with, followed by filtration to remove AgCl and by evaporation of the solvent, was heated at 100°C overnight under vacuum. Mixtures containing up to 70% of **18** (by ES/MS and HPLC) together with the starting CB₁₁Me₁₂⁻ were prepared this way. However, this procedure was not used for preparative purposes because of the difficulties with separating **18** from the starting **2**.

Cs⁺ 1–Cl(CH₂)₆–CB₁₁H₁₁⁻ (27)

Cs⁺ 1-Cl(CH$_2$)$_6$-CB$_{11}$H$_{11}^-$ (27)

Compound **27** was prepared as a white solid from 1.02 g (5 mmol) of Me₃NH⁺CB₁₁H₁₂⁻ by reaction with 3.0 mL of 1.7 M *n*-BuLi (5.1 mmol) in THF followed by 0.74 mL of 1,6–dichlorohexane (0.79 g, 5.1 mmol) in 86% yield (1.73 g). It was purified by crystallization from water. ¹H {¹¹B} NMR (acetone-d_6): δ 3.57 (2 H, t, 6'-CH₂), 1.75 (2 H, m, 5'-CH₂), 1.51 (15 H, s, 2–6–B), 1.46 (15 H, s, 7–11–B), 1.31 (4 H, m, 4',3'-CH₂), 1.16 (2 H, m, 2'-CH₂), 0.789 (2 H, t, 1'-CH₂). ¹¹B {¹H} NMR (acetone-d_6): δ -9.73 (B₁₂), –13.40 (B$_{2-11}$). ¹³C {¹H} NMR (acetone-d_6): δ 59.13 (6'-CH₂), 53.95 (1–C), 43.16 (5'-CH₂), 33.85

(4'-CH$_2$), 29.24 (3'-CH$_3$), 23.40 (2'-CH$_2$), 14.67 (1'-CH$_2$). IR (KBr pellet): 715, 949, 1033, 1191, 1285, 1317, 1387, 1455, 2545, 2865, 2956 cm^{-1}. MS/ES(-) in methanol: base peak at m/e 261 with the expected isotopic distribution (M⁻). Anal. Calcd for C$_7$H$_{23}$B$_{11}$ClCs: C, 21.31; H, 5.88. Found: C, 21.59; H, 5.67.

Cs⁺ 1–Cl(CH$_2$)$_6$–CB$_{11}$Me$_{11}$⁻ (28)

Compound **28** was prepared from 1.18 g (3 mmol) of **32**, 10.4 mL (92 mmol) of MeOTf and 10.4 g (247 mmol) of CaH$_2$ using a reported [] permethylation procedure. Yield 1.51 g (92%). It was purified by crystallization from water. ^{1}H NMR (acetone-d_6): δ 3.56 (2 H, t, 6'-CH$_2$), 1.79 (2 H, m, 5'-CH$_2$), 1.31 (4 H, m, 4',3'-CH$_2$), 1.16 (2 H, m, 2'-CH$_2$), 0.795 (2 H, t, 1'-CH$_2$)-0.210 (15 H, s, 7–11–CH$_3$), –0.435 (15 H, s, 2–6–CH$_3$), –0.489 (3 H, s, 12–CH$_3$). ^{11}B {^{1}H} NMR (acetone-d_6): δ -0.44 (B$_{12}$), –9.14 (B$_{7-11}$), –10.72 (B$_{2-6}$). ^{13}C {^{1}H} NMR (acetone-d_6): δ 58.73 (6'-CH$_2$), 50.63 (1–C), 43.26 (5'-CH$_2$), 32.14 (4'-CH$_2$), 29.44 (3'-CH$_3$), 23.87 (2'-CH$_2$), 14.76 (1'-CH$_2$), –2.84 (2–12–CH$_3$). IR (KBr pellet): 710, 909, 1152, 1279, 1311, 1429, 2827, 2854, 2891, 2928 cm^{-1}. MS/ES(-) in methanol: base peak at m/e 415 with the expected isotopic distribution (M⁻). Anal. Calcd for C$_{18}$H$_{45}$B$_{11}$ClCs: C, 39.39; H, 8.26. Found: C, 39.68; H, 7.89.

Attempted Elimination Reactions for 28

Several elimination conditions have been tested on a small scale (22.3 mg, 0.05 mmol of **28**). Using an excess of KOEt in EtOH at 80°C for 2 days resulted (by ES/MS) in a mixture ca. 30% of the starting material (m/e 415) and an ethoxy substituted product (m/e 426), with a trace amount of the desired alkene (m/e 379). Using an excess of potassium ethylene glycolate at room temperature for 2 days resulted (by ES/MS) in no reaction, while heating this reaction mixture at 80°C for 2 days resulted in almost complete conversion into the corresponding ethylene glycol ether (m/e 441) with a trace amount of the desired alkene. Using an excess of t-BuONa in t-BuOH at 80°C for 2 days gave no reaction. Excess CsF in DMF at 105°C for 2 days provided (by ES/MS) a mixture of ca. 40% of the desired alkene (m/e 379), 40% of a fluorine substituted product (m/e 399), and 20% of the starting material (m/e 415).

Reactions of 18 with Benzyl Chloride and 4–Vinylbenzyl Chloride

When 22.3 mg (0.05 mmol) of **18** were treated with 0.06 mL (0.1 mmol) of n-BuLi (1.7 M solution in hexanes) in THF for 2 h at r.t., followed by 11.5 μL of benzyl chloride (12.7 mg, 0.1 mmol), only the starting material was found (ES/MS) in the reaction mixture after hydrolysis. When the same amount of **18** was reacted with benzyl chloride in 3 mL of THF in the presence of an excess NaH and NaI, a mixture containing ca. 80% of the benzyl substituted product (m/e 403) and 20% of the starting material was found by ES/MS. However, when the same conditions were used with 4–vinylbenzyl chloride a complicated mixture of carborane containing products, none of which corresponded to the desired product (m/e 429), was found by ES/MS. The reaction with 4–vinylbenzyl chloride (15.2 mg, 0.1 mmol)) was also repeated using the 12–OLi-**18**, generated by reacting **18** (22.3 mg, 0.05 mmol) with 60 μL (0.1 mmol) of 1.6 M n-BuLi, under reflux in THF. Only the starting material was found by ES/MS in the reaction mixture after hydrolysis.

Reactions of 18 Poly(vinylbenzyl chloride) and Merrifield Resin

To a solution of 23.3 mg (0.05 mmol) of **18** in 5 mL of dry THF were added 40 μL (0.06 mmol) of 1.5 M solution of MeLi in ether, followed by 70 mg of poly(vinylbenzyl chloride). The reaction mixture was refluxed and monitored by NMR. No change in the NMR was observed after several days, and the starting material (22 mg, 94%) was recovered from the reaction mixture after hydrolysis. The same result was obtained when Merrifield resin was used instead of poly(vinylbenzyl chloride).

Reactions of 18 with 5–Bromo-1–Pentene

No reaction was observed (ES/MS) when **18** (22.3 mg, 0.05 mmol) was treated with 59.2 μL (74.5 mg, 0.5 mmol) of 5–bromo-1–pentene inTHF in the presence of an excess of NaH and NaI. When 22.3 mg (0.05 mmol) of **18** were treated with 60 μL (0.1 mmol) of n-BuLi (1.7 M solution in hexanes) in THF for 2 h at room temperature, followed by 12 μL (15 mg, 0.1 mmol) of 5–bromo-1–pentene, only the starting material was found (ES/MS) in the reaction mixture after hydrolysis.

Reaction of 18 with Pentenyl Triflate

10 mg (0.022 mmol) of **18** were stirred with 90 mg (0.4 mmol) of 1–pentenyl triflate in 2 mL of sulfolane in the presence of 0.2 g of CaH_2. The reaction was monitored by ES/MS and after 24 h at r.t. only the desired product (m/e 381) was found in the reaction mixture.

Reactions of 18 Reactions of Li^+ $CB_{11}Me_{12}^-$ with Polystyrene and Poly(phenylmethylsiloxane)

The lithium salt of $CB_{11}Me_{12}^-$ (40 mg) was dried under vacuum at 150°C overnight in 5 mL ampules. 200 mg of polystyrene or poly(phenyl-methylsiloxane) were added to the ampules under Ar, ampules were evacuated, sealed, and heated in an oil bath at 160°C for three days. During this time the lithium salt dissolved in the polymers almost completely and solutions became yellow. These solutions were used directly for the conductivity measurements.

ACKNOWLEGMENTS

I would like to thank Professor Josef Michl for his invaluable support and encouragement. I would also like to thank Dr. David F. Pickett (AAAA Energy Enterprises, Inc.) for the collaboration. I am grateful to the Link Foundation for the generous support of this research, which was also partly supported by US Air Force STTR Program (F33615–99–C-2980).

REFERENCES

[1] C. Julien. "Solid State Batteries" In *CRC Handbook of Solid State Electrochemistry*; P. J. Gellings and H. J. M. Bouwmeester, Eds.; CRC: Boca Raton, 371–406 (1997).

[2] *Lithium Ion Batteries*. M. Wakihara and O. Yamamoto, Eds.; Wiley-VCH: Berlin, New York, Chichester, Brisbane, Singapore, Toronto (1998).

[3] D. Fauteux, A. Massucco; . McLin, M. van Buren and J. Shi, "Lithium Polymer Electrolyte Rechargeable Battery" *Electrochim. Acta* 40, 2185–2190 (1995).

[4] (a) V. S. Bagotsky and A. M. Skundin, "Fundamental Scientific Problems Related to the Development of Rechargeable Lithium Batteries" *Russ. J. Electrochem.* 37(7), 654–661 (1998). (b) *Polymer Electrolytes Review.* J. R. McCallum and C. A. Vincent, Eds.; Elsevier Applied Science: London. Vol. 1, 1987; Vol. 2 (1989).

[5] A. G. Einset and G. E. Wnek, "Polymer Electrolyte Review" In *Handbook of Solid State Batteries & Capacitors;* M. Z. A. Munshi, Ed.; World Scientific: Singapore, New Jersey, London, Hong Kong, 289–309 (1995).

[6] (a) P. V Wright, "Polymer Electrolytes—the Early Days" *Electrochim. Acta* 43(10–11), 1137–1143 (1998). (b) M. Z. A. Munshi, "Technology Assessment of Lithium Polymer Electrolyte Secondary Batteries" In *Handbook of Solid State Batteries & Capacitors;*M. Z. A. Munshi, Ed.; World Scientific: Singapore, New Jersey, London, Hong Kong, 393–423 (1995).

[7] B. E. Fenton, J. M. Parker, and P. V. Wright, "Complexes of Alkali Metal Ions with Poly(ethylene oxide)" *Polymer* 14, 589–595 (1973).

[8] M. B. Armand, "Polymer Electrolytes" *Annu. Rev. Mater. Sci.* 16, 245–261 (1986).

[9] (a) M. Stainer, L. C. Hardy, D. H. Whitmore, and D. F. Schriver, "Stoichiometry of Formation and Conductivity of Amorphous and Crystalline Complexes Formed between Poly(ethylene oxide) and Ammonium Salts: $PEO_x NH_4 SCN$ and $PEO_x NH_4 SO_3 CF_3$ " *J. Electrochem. Soc.* 131, 784–790 (1984). (b) P. Ferloni, G. Chiodelli, A. Magistris, and M. Sanesi, "Ion Transport and Thermal Properties of Poly(ethylene oxide)—Lithium Perchlorate Polymer Electrolytes" *Solid*

State Ionics 18/19, 265–270 (1986). (c) J. E. Weston and B. C. H. Steele, "Thermal History-Conductivity Relationship in the Lithium Salt—Poly(ethylene oxide) Complex Polymer Electrolytes" *Solid State Ionics* 2, 347–354 (1981). (d) M. Minier, C. Berthier, and W. Gorecki, "Differential Scanning Calorimetry of Soldium Iodide Polyethylene Oxide Complexes" *Solid State Ionics* 9/10, 1125–1127 (1983). (e) P. R. Soerensen and T. Jacobsen, "Phase Diagram and Conductivity of the Polymer Electrolyte PEORLiCF$_3$SO$_3$" *Polymer Bull.* 9, 47–51 (1983). (f) R. Dupon, B. L. Papke. M. A. Ratner, D. H. Whitmore, and D. F. Schriver, "Influenece of Ion Pairing on Cation Transport in the Polymer Electrolytes Formed by Poly(ethylene oxide) with Sodium Tetrafluoroborate and Sodium Tetrahydroborate" *J. Am. Chem. Soc.* 104, 6247–6251 (1982).

[10] E. Quartone, P. Mustarelli, and A. Magistri, "PEO-Based Composite Polymer Electrolytes" *Solid State Ionics* 110(1–2), 1–14 (1998).

[11] J. R. M. Giles, "Electrolytic Conduction in Amorphous Salt Complexed Polyethers" *Solid State Ionics* 24, 155–167 (1987).

[12] J. R. M. Giles, F. M. Gray, J. R. McCallum, and C. A. Vincent, "Synthesis and Characterization of ABA Block Copolymer-Based Polymer Electrolytes" *Polymer* 28, 1977–1981 (1987).

[13] (a) J. Sun, D. R. McFarlane, and M. Forsyth, "Mechanical Properties of Polyether—Plasticizer—Salt Systems as Polymer Electrolytes" *Solid State Ionics* 85, 137–141 (1996). (b) I. Albinsson, B.-E. Mellander, and J. R. Stevens, "Ionic Conductivity in Poly(ethylene oxide) Modified Poly(dimerthylsiloxane) Complexed with Lithium Salts" *Polymer* 32, 2712–2715 (1991).

[14] J. R. M. Giles, and M. P. Greenhall, "Ionic Conduction in Phosphate Ester-Crosslinked Polyethylene Glycols Complexed with Lithium Trifluoromethanesulfonate" *Polymer Comm.* 27, 360–362 (1986).

[15] (a) G. C. Rawsky, T. Fujinami, and D. F. Shriver, "Aluminosilicate/Polyethylene Glycol Copolymers: A New Class of Polyelectrolytes" *Polym. Mater. Sci. Eng.* 71, 523–524 (1994). (b) H. Liu, Y. Okamoto, and T. A. Skotheim, "Fluorenyllithium Salt Containing Polymers as Single-Ion Conductors" *Mol. Cryst. Liq. Cryst.* 190, 213–220 (1990).

[16] W. H. Meyer, "Polymer Electrolytes for Lithium-Ion Batteries" *Adv. Mater.* 10(6), 439–448 (1998).

[17] C. A. Angell, K. Xu, S.-S. Zhang, and M. Videa, "Variations on the Salt-Polymer Electrolyte Theme for Flexible Solid Electrolytes" *Solid State Ionics* 86–88, 17–28 (1996).

[18] (a) G. Ardel, D. Golodnitsky, E. Peled, Y. Wang, S. Bajue, and S. Greenbaum, "Bulk and Interfacial Ionic Conduction in LiI/Al$_2$O$_3$ Mixtures" *Solid State Ionics* 113–115, 477–485 (1998). (b) G. Nagasubramanian, E. Peled, A. I. Attia, and G. Halpert, "Composite Solid Electrolyte for Li Battery Applications" *Proc. Electrochem. Soc.* 93, 86–97 (1993).

[19] (a) N. Ogata, K. Sanui, M. Rikukawa, W. Yamada, and M. Watanabe, "Super Ion Conducting Polymers for Solid Polymer Electrolytes" *Synth. Met.* 69 (1–3), 521–524 (1995). (b) M. Watanabe, S.-I. Yamada, K. Sanui, and N. Ogata, "High Ionic Conductivity of New Polymer Electrolytes Consisting of Polypyridinium, Pyridinium and Alumiunm Chloride" *J. Chem. Soc., Chem. Comm.* 929–931 (1993).

[20] B. Scrosati, "Conducting Polymers: Advanced Materials for New Design, Rechargeable Lithium Batteries" *Polym. Int.* 47 (1), 50–55 (1998).

[21] N. J. Dudney, "Composite Electrolytes" In *Handbook of Solid State Batteries & Capacitors*;M. Z. A. Munshi, Ed.; World Scientific: Singapore, New Jersey, London, Hong Kong, 231–246 (1995).

[22] (a) F. M. Gray, *Polymer Electrolytes*; Royal Soc. Chem., UK; 1997. (b) F. M. Gray, *Solid Polymer Electrolytes. Fundumentals and Technological Applications*; VCH Publishers (1991).

[23] (a) R. N. Grimes, *Carboranes*; Academic Press: New York and London, 1970. (b) T. Onak, "Organoboron Chemistry" In *Comprenhensive Organometallic Chemistry II*; E. W. Abel, F. G. Stone, and G. Wilkinson, Eds.; Elsevier Science Ltd.: Oxford, Vol. 1, Ch. 6 (1995).

[24] (a) W. H. Knoth, "1–B_9H_9CH- and $B_{11}H_{11}CH$-" *J. Am. Chem. Soc.* 89, 1274–1275 (1967). (b) J. Plešek, T.; Jelínek, E. Drdáková, S. He mánek, and B. Štíbr, "A Convenient Preparation of Dodecahydro-1–carba-*closo*-dodecaborane (1–) ion and its C-Amino Derivatives" *Collect. Czech. Chem. Comm.* 49, 1559–1562 (1984).

[25] B. T. King, Z. Janoušek, B. Gr ner, M. Trammell, B. C. Noll, and J. Michl, "Dodecamethyl-carba-*closo*-dodecaborate (1–) Anion, $CB_{11}Me_{12}^-$" *J. Am. Chem. Soc.* 118, 3313–3314 (1996).

[26] S. Moss, B. T. King, and J. Michl, unpublished results.

[27] B. T. King, B. C. Noll, A. J. McKinley, and J. Michl,. "Dodecamethylcarba-*closo*-dodecaboranyl ($CB_{11}Me_{12}$), a Stable Free Radical" *J. Am. Chem. Soc.* 118, 10902–10903 (1996).

[28] S. H. Strauss, "The Search for Larger and More Weakly Coordinating Anions" *Chem. Rev.* 927–942 (1993).

[29] I. Zharov, S. Moss, and J. Michl, unpublished results.

[30] N. Ogata, *"Ion-Conducting Polymers"* In *Functional Monomers and Polymers*. K. Takemoto, R. M. Ottenbrite, and M. Kamachi, Eds.; Dekker: New York, 387–431 (1997).

[31] P. V. Wright, Y. Zheng, D. Bhatt, T. Richardson, and G. Ungar, "Supramolecular Order in New Polymer Electrolytes" *Polym. Int.* 47(1), 34–42 (1998).

[32] M. B. Armand, W. Gorecki, and R. Andreani, "Perfluorosulfonimide Salts as Solute for Polymer Electrolytes" In *Proc. 2nd Int. Symp. on Polymer Electrolytes*. B. Scrosati, Ed.; Elsevier: New York, 91–97 (1990).

[33] M. F.; Gauthier, M. Armand, D. Muller, "New Electrolyte Solutions for Batteries" In *Electroresponsive Molecules and Polymeric Systems*. T. A. Skotheim, Ed.; Dekker: New York, Vol. 1, 245–259 (1998).

[34] T. Jelínek, P. Baldwin, W. R. Scheidt, and C. A. Reed, "New Weakly Coordinating Anions. 2. Derivatization of the Carborane Anion $CB_{11}H_{12}^-$" *Inorg. Chem.* 32, 1982–1990 (1993).

[35] T. Jelínek, J. Plešek, S. He mánek, and B. Štíbr, "Chemistry of Compounds with the 1–Carba-*closo*-dodecaborane(12) Framework" *Collect. Czech. Chem. Commun.* 51, 819–829 (1986).

[36] B. Gr ner, Z. Janoušek, B. T. King, J. N. Woodford, C. H. Wang, V. Všete ka, and J. Michl, "Synthesis of 12–Substituted 1–Carba-*closo*-dodecaborate Anions and First Hyperpolarizability of the 12–$C_7H_6^+$-$CB_{11}H_{11}^-$ Ylide" *J. Am. Chem. Soc.* 121, 3122–3126 (1999).

[37] J. T. Ceremuga, B. T. King, J. R. Clayton, and J. Michl, unpublished results.
[38] Z. Xie, J. Manning, R. Reed, R. Mathur, P. D. W. Boyd, A. Benesi, and C. A. Reed, "Approaching the Silylium (R_3Si^+) Ion: Trends with Hexahalo (Cl, Br, I) Carboranes as Counterions" *J. Am. Chem. Soc.* 118, 2922–2928 (1996).
[39] I. Zharov, B. T. King, and J. Michl, unpublished results.
[40] B. T. King and J. Michl. *J. Am. Chem. Soc.* accepted for publication.
[41] S. M. Pillai, M. Ravindranathan, and S. Sivaram, "Dimerization of Ethylene and Propylene Catalyzed by Transition-Metal Complexes" *Chem. Rev.* 86, 353–399 (1986).
[42] W. Noll, *Chemistry and Technology of Silicones.* Academic Press Inc.: New York, London (1968).
[43] J. R. Clayton, M. Sc. Thesis, University of Colorado, Boulder (1999).
[44] P. A. Deck and T. J. Marks, "Cationic Metallocene Olefin Polymerization Catalysts. Thermodynamic and Kinetic Parameters for Ion Pair Formation, Dissociation, and Reorganization" *J. Am. Chem. Soc.* 117, 6128–6129 (1995).
[45] B. L. Small, M. Brookhart, and A. M. A. Bennett "Highly Active Iron and Cobalt Catalysts for the Polymerization of Ethylene" *J. Am. Chem. Soc.* 120, 4049–4050 (1998).

Supramolecular Solar Energy Conversion with Nanocrystalline TiO$_2$

Georg M. Hasselmann

Department of Chemistry, Johns Hopkins University

Baltimore, Maryland 21218

Research Advisor: Dr. Gerald J. Meyer

ABSTRACT

The sensitizer [Ru(dcb)$_2$(Cl)-bpa-Os(bpy)$_2$(Cl)](PF$_6$)$_2$, abbreviated Ru-bpa-Os ,where dcb is 4,4'-(CO$_2$H)$_2$-2,2'-bipyridine and bpa is 1,2–bis(4–pyridyl)ethane, was prepared, characterized, and anchored to nano-crystalline (anatase) TiO$_2$ films for interfacial electron transfer studies. 417 nm or 532 nm light excitation of a TiO$_2$ | Ru-bpa-Os material immersed in a 1.0 M LiClO$_4$ acetonitrile bath at 25°C results in rapid interfacial electron transfer and intramolecular electron transfer to ultimately form an interfacial charge separated state with an electron in TiO$_2$ and an oxidized Os(III) center, abbreviated TiO$_2$(e$^-$) | Ru-bpa-Os(III). This same state can also be generated after selective excitation of the Os(II) moiety with 683 nm light. The rates of intramolecular and interfacial electron transfer are fast, $k > 10^8$ s^{-1}, while interfacial charge recombination, TiO$_2$(e$^-$) | Ru-bpa-Os(III) $\rightarrow$ TiO$_2$ | Ru-bpa-Os, requires milliseconds for completion.

INTRODUCTION

Supramolecular coordination compounds represent a new class of chromophores for sensitization of wide bandgap semiconductors to visible light [1–7]. An important strategy for this application is shown in Scheme 1 where a sensitizer, S, is anchored to a semiconductor surface, SC, with an electron donor, D, covalently bound through a bridging ligand, L [6,7]. Light excitation of the sensitizer forms an excited state, S*, that rapidly injects an electron into the semiconductor, Step 1. Intramolecular electron transfer translates the "hole" from the oxidized sensitizer to the covalently bound donor, Step 2. This

Scheme 1

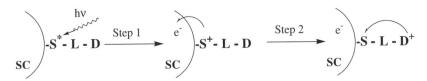

sequence of electron transfer can result in enhanced lifetime of the interfacial charge separated state, improved power output in regenerative solar cells, and photochromic materials [6,7]. Inorganic donors provide a high degree of molecular flexibility as systematic changes in the ancillary ligands can be used to finely tune the driving force for intramolecular electron transfer, step 2, and the spectral properties of the materials. Here we report the first proof-of-concept example with semiconductor-bound bimetallic coordination compounds that behave like molecular photodiodes: light promotes rapid electron transfer into and hole transfer away from the solid.

MATERIALS AND METHODS

Preparation and Characterization of the Compounds:

The complexes Ru(dcH$_2$b)$_2$(Cl)$_2$ and Os(bpy)$_2$(Cl)$_2$ were available from previous studies.

[Os(bpy)$_2$(Cl)bpa](PF$_6$) A 0.2 g amount of Os(bpy)$_2$(Cl)$_2$ and 1 g of bpa were dissolved in 180 ml ethanol/water (1:1, v:v) and refluxed for 2 h under Ar. The solution was rotary evaporated to a few ml and cooled. The precipitated bpa ligand was filtered off and the crude product was precipitated by addition of NH$_4$PF$_6$. The product was dissolved in a minimum volume of acetonitrile and reprecipitated with diethylether after addition of a few drops of hydrated hydrazine. The product was further purified by chromatography on alumina eluting with acetonitrile/toluene (1:1, v:v). The complex was obtained as a brown solid (0.18 g, 60%). Elemental analysis for $C_{32}H_{28}N_6ClOsPF_6$: C 44.32 (43.85), H 3.25 (3.12), and N 9.69 (9.48) calc. (found).

Ru(dcH$_2$b)(dcHb)(Cl)py A 0.15 g amount of Ru(dcH$_2$b)(Cl)$_2$ was converted into the corresponding TBA$_4$[Ru(dcb)$_2$(Cl)$_2$] by titration with tetrabutylammonium hydroxide. To a methanol solution of the salt, 20 l of pyridine was added and the mixture refluxed for 2 h. After evaporation to dryness, the complex was dissolved in water, precipitated at pH 3 by addition of 2 M HCl and dried under vacuum. Elemental analysis for $C_{29}H_{20}N_5RuClO_8$: C 44.94 (43.85), H 3.64 (3.58), and N 9.04 (9.11) calc. (found).

[(Cl)Os(bpy)$_2$bpaRu(dcH$_2$b)$_2$(Cl)](PF$_6$)$_2$ A 0.165 g amount of [Os(bpy)$_2$ (Cl)bpa](PF$_6$) and 0.114 g of K$_4$Ru(dcb)(Cl)$_2$ were refluxed in 80 ml of ethanol/water (1:1, v:v) for 3 hours under Ar. The solution was rotary evaporated to ca. 5 ml under reduced light and chromatographed on Sephadex G15 eluting with 0.01 M NaCl. HPF$_6$ was added to the first eluted brown fraction and the solid filtered, dried under vacuum and recrystallized from methanol/diethylether. Elemental analysis for $C_{56}H_{44}O_8N_{10}RuOsP_2F_{12}$: C 41.09 (40.85), H 2.71 (2.64), and N 8.56 (8.51) calc. (found).

The coordination compounds anchor to nanocrystalline TiO$_2$ films in high surface coverages as previously described [8].

Spectroscopic Measurements

Absorbance: UV-Vis absorbance measurements were made on a Hewlett-Packard 8453 Diode Array Spectrophotometer.

Transient Absorption: Measurements were carried out on the apparatus previously described [8]. Briefly, excitation was carried out using the 532 nm laser pulses, ca. 8 ns and 10 mJ/pulse, from a Nd:YAG (Continuum Surelite). The approximately 5 mm diameter excitation beam was expanded to ca. 3 cm using a quartz plano-concave lens (JML Direct, −50 mm EFL, 25.4 mm diameter), resulting in a fluence of around 3 mJ cm^{-2}. The absorbance change of the laser irradiated sample was probed at 90° to the excitation pulse using an Applied Photophysics 150 W Xe arc lamp operating in pulsed mode. The sample was protected from the probe light using a fast shutter, 10 ms pulse width, and appropriate UV and heat absorbing glass and solution filter combinations. The probe light was focused onto the sample and again onto the entrance slit of a f/3.4 Applied Photophysics monochromator, typically under conditions such that the effective bandwidth was 2–3 nm. The probe beam was monitored after the monochromator using a Hamamatsu R928 photomultiplier. The photomultiplier was protected from scattered laser light using appropriate glass filters positioned between the sample and monochromator. Kinetic traces were recorded on a LeCroy 9450 digital oscilloscope using a 50 Ω input and operating at 350 MHz. In general, kinetic traces represent the average of 40–100 laser shots. Excitation was directed to the front face of the TiO$_2$ film, oriented at a ca. 45° angle, such that the predominant Raleigh scattering was directed away from the monochromator. Cuvettes were stoppered with a PTFE plug, argon purged by bubbling through a glass tube and maintained under a premoistened argon flow.

Photoluminescence. Photoluminescence (PL) spectra were obtained with a Spex Fluorolog which had been calibrated with a standard NBS tungsten-halogen lamp. Emission lifetimes were measured with a nitrogen laser pumped dye-laser setup. The emission was collected at a right angle with a Sciencetech $f/3.4$ monochromator equipped with a Hamamatsu R928 photomultiplier tube.

Electrochemical Measurements:

Electrochemistry. Electrochemistry was performed in 0.1 M tetrabutylammonium hexafluorophosphate (Aldrich, TBAH) methanol or acetonitrile electrolyte. The TBAH was recrystallized from ethanol. A BAS Model CV-50W potentiostat was used in a standard three cell arrangement consisting of a Pt working electrode, a Pt gauze counter electrode, and a SCE reference electrode. Approximately mM concentrations of the compounds were dissolved in the electrolyte.

Photoelectrochemistry. Photoelectrochemical measurements were performed in a 2 electrode sandwich cell arrangement as previously described [8]. Briefly, ~ 10 μl of electrolyte was sandwiched between a TiO2 electrode and a Pt coated tin oxide electrode. The TiO2 was illuminated with a 450 W Xe lamp coupled to a $f/0.22$ m monochromator for IPCE measurements. Light excitation was through the FTO glass substrate of the photoanode. Photocurrents were measured under short circuit conditions with a Keithly Model 617 digital electrometer. Incident irradiances were measured with a calibrated silicon photodiode from UDT Technologies. The supporting electrolyte was 0.5 M LiI/0.05 M I2 in acetonitrile.

RESULTS AND DISCUSSION

The compound of interest is [Ru(dcb)$_2$(Cl)-bpa-Os(bpy)$_2$(Cl)](PF$_6$)$_2$, abbreviated Ru-bpa-Os where dcb is 4,4'-(COOH)$_2$–2,2'-bipyridine and bpa is 1,2–bis(4–pyridyl)ethane. An idealized structure of this coordination compound bound to TiO_2, abbreviated TiO_2 I Ru-bpa-Os, is shown in Scheme 2.

Scheme 2

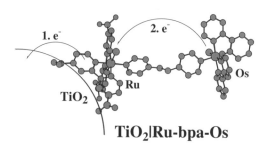

TiO_2IRu-bpa-Os

The bpa bridge provides weak electronic coupling between Ru(II) and Os(II) as shown by the lack of a measurable intervalence charge transfer band in the one electron oxidized form [9]. The visible absorption spectrum is well represented as a sum of the individual $Ru(dcb)_2(py)Cl$ and $Os(bpy)_2(bpa)Cl^+$ spectra with overlapping Ru → dcb and Os → bpy charge transfer bands centers ~ 450 nm and a spin forbidden ^3MLCT, Os → bpy, charge transfer band centered around 720 nm, Figure 1. The bimetallic and model compounds are non-emissive in acetonitrile or methanol solutions at room temperature.

Cyclic voltammetry of the bimetallic compound in 1.0 M $LiClO_4$ yields $E_{1/2}(Ru(III/II))$ = + 0.98 V and $E_{1/2}(Os(III/II))$ = + 0.36 V vs. SCE, Figure 2. The model compound $Ru(dcb)_2(py)(Cl)$ gives a single wave with $E_{1/2}(Ru(III/II))$ = + 0.9 V vs SCE in methanol electrolyte solution.

Pulsed blue (417 nm), green (532 nm), or red (683 nm) light excitation of TiO_2 I Ru-bpa-Os result in indistinguishable absorption difference spectra, Figure 3.

Comparison of this spectra with one generated by stoichiometric oxidation of Os(II) with Ce(IV) show that the transient is reasonably assigned to an interfacial charge separated state with an electron in TiO_2 and an Os(III) center, abbreviated $TiO_2(e^-)$ I Ru-bpa-Os(III). Red light (683 nm) selectively forms the Os → bpy metal-to-ligand charge transfer (MLCT) excited state, that rapidly inject an electron into TiO_2 with an injection rate constant(s), k_{inj}, faster then our instrument response, Equation 1. Remote injection of this type probably results from the

$$(1)$$

flexible bpa bridge that allows alternative surface orientations, not shown in the idealized Scheme 2, which park the Os(II) centers proximate to the semiconductor surface [5].

Blue (417 nm) or green (532 nm) light excitation produce both the Ru

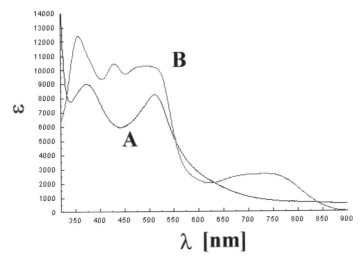

Figure 1. Absorption spectra of Ru(dcb)$_2$(py)(Cl) (A) and Os(bpy)$_2$(bpa)Cl$^+$ (B) in acetonitrile.

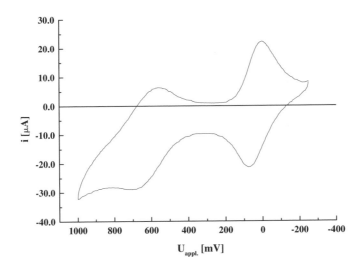

Figure 2. Cyclic voltammogram of Ru-bpa-Os adsorbed on FTO glass (working electrode). The data was acquired at a 200 mV/s scan rate, 0.1 M TBAPF$_6$/acetonitrile electrolyte, Pt gauze as auxiliary electrode, and Ag/AgNO$_3$ in acetonitrile as a non-aqueous reference electrode.

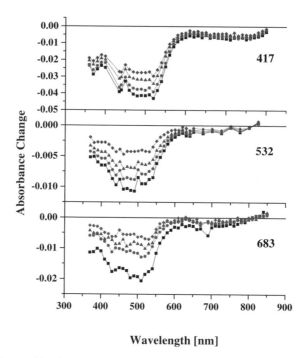

Wavelength [nm]

Figure 3. The visible absorption difference spectrum of a TiO_2 | Ru-bpa-Os material immersed in a 1.0 M $LiClO_4$ acetonitrile solution at 25°C recorded 0 ns (squares), 100 ns (circles), 250 ns (triangles), and 500 ns (diamonds) after pulsed (8–10 ns fwhm, ~ 5 mJ/cm²) light excitation. The excitation wavelength was 417 nm, 532 nm, and 683 nm, respectively. The normalized spectra (not shown) are within experimental error the same and assigned to an interfacial charge separated state with an electron in TiO_2 and an Os(III) center, abbreviated $TiO_2(e^-)$ | Ru-bpa-Os(III).

and Os MLCT excited states in relative concentrations proportional to their ground state absorption at these wavelengths. We have demonstrated that light absorbed by the Ru(II) chromophore is converted into charge separated states by comparative actinometry studies of TiO_2 | Ru(dcb)₂(py)Cl and TiO_2 | Ru-bpa-Os. The injection yields in 1.0 M $LiClO_4$ in acetonitrile with 532 nm excitation were the same for these two sensitized materials within experimental error, ±10%. Therefore, Ru(II)* excited states lead to charge separation most probably by direct injection into TiO_2, Equation 2 [10].

$$\cdot \qquad (2)$$

The product of Equation 2, $TiO_2(e^-)$ | Ru^{III}-bpa-Os^{II} is not observed spectroscopically under any conditions due to rapid intramolecular hole transfer from Ru(III) to Os(II) , Equation 3.

$$TiO_2(e^-)\left|Ru^{III} - bpa - Os^{II} \xrightarrow{k_{et(2)}>10^8\,s^{-1}} TiO_2(e^-)\right|Ru^{II} - bpa - Os^{III} \cdot \quad (3)$$

The driving force for intramolecular hole transfer is 0.6 V and $k_{et(2)} > 10^8$ s^{-1}. Interfacial charge recombination of the electron in TiO$_2$ with the Os(III) center, Equation 4,

$$TiO_2(e^-)\left|Ru^{II} - bpa - Os^{III} \xrightarrow{k_{cr}} TiO_2\right|Ru^{II} - bpa - Os^{II} \cdot \quad (4)$$

requires milliseconds for completion. The interfacial electron transfer rates are independent of whether the TiO$_2$(e$^-$) | Ru-bpa-Os(III) state was created by direct injection from Os (II), red light excitation, or by the two-step path depicted in Equations 3 and 4. Typical charge recombination data is given in Figures 4 and 5.

Figure 1 shows the absorption spectra of Ru(dcb)$_2$(py)Cl, and Os(bpy)$_2$(bpa)Cl$^+$, the sum of which corresponds to the bimetallic compound [Ru(dcb)$_2$(Cl)-bpa-Os(bpy)$_2$(Cl)](PF$_6$)$_2$. With 532 nm light excitation of [Ru(dcb)$_2$(Cl)-bpa-Os(bpy)$_2$(Cl)](PF$_6$)$_2$, approximately 55% of the light is

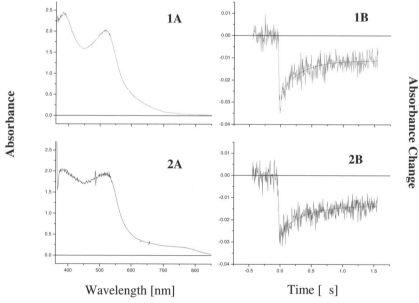

Figure 4. Absorption of TiO$_2$ | Ru(dcb)$_2$(py)Cl (1A) and TiO$_2$ | Ru(dcb)$_2$(Cl)-bpa-Os(bpy)$_2$(Cl) (2A) in acetonitrile measured vs. a blank TiO$_2$ film reference. The panels on the right show the recovery of the bleach of TiO$_2$ | Ru(dcb)$_2$(py)Cl (1B) and TiO$_2$ | Ru(dcb)$_2$(Cl)-bpa-Os(bpy)$_2$(Cl) (2B) in 1.0 M LiClO$_4$/acetonitrile at 500 nm after excitation with 532 nm laser pulse (8–10 ns fwhm, ~5 mJ/cm^2). 40 kinetic traces are averaged.

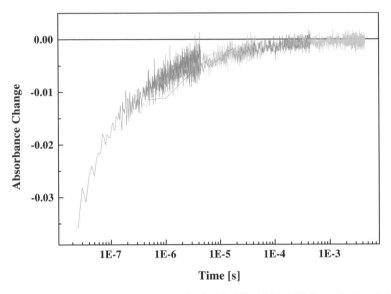

Figure 5. Transient absorption signal of TiO_2 I $Ru(dcb)_2(Cl)$-bpa-$Os(bpy)_2(Cl)$ monitored at 500 nm after excitation with 532 nm at different oscilloscope digitizing rates. 160 shots per data set averaged.

absorbed by the Os(II) chromophore and ca. 45% by Ru(II). The relative injection yields of TiO_2 I $Ru(dcb)_2(py)Cl$ and TiO_2 I $Ru(dcb)_2(Cl)$-bpa-$Os(bpy)_2(Cl)$ were measured by comparative actinometry with 532 nm excitation in 1.0 M $LiClO_4$/acetonitrile solution under conditions where the ground state absorption at the excitation wavelength are approximately the same. Ten nanoseconds after light excitation of TiO_2 I $Ru(dcb)_2(py)Cl$ the charge separated state TiO_2 I $Ru^{III}(dcb)_2(py)Cl$ is formed, while excitation of TiO_2 I $Ru(dcb)_2(Cl)$-bpa-$Os(bpy)_2(Cl)$ forms a TiO_2 I $Ru(dcb)_2(Cl)$-bpa-$Os^{III}(bpy)_2(Cl)$ interfacial charge separated state. Since the excited state is very short lived, $\tau \ll 10$ ns, and the oxidized forms of the compounds do not absorb appreciably in the visible region, the magnitude of the absorbance transient is directly related to the injection yield. There may be a minor contribution from the injected electron at long wavelengths.

Shown in Figure 4 are the ground state absorption spectra of two sensitized films TiO_2 I $Ru(dcb)_2(py)Cl$ (1A) and TiO_2 I $Ru(dcb)_2(Cl)$-bpa-$Os(bpy)_2(Cl)$ (2A) respectively. The corresponding absorption transients measured at 500 nm are shown (1B and 2B). Superimposed on the data is a second-order kinetic model plus a non-zero baseline. The time zero amplitudes are within experimental error the same indicating that the absorbed light is converted to interfacial charge separated states with the same relative efficiency. Furthermore, actinometry with optically matched

Ru(dcb)(bpy)$_2$$^{2+}$ | TiO$_2$ samples, which has an injection yield of unity (φ_{inj} = 1.0 ± 0.1) under these conditions [11] demonstrates that φ_{inj} = 1 within experimental error for both TiO$_2$ | Ru(dcb)$_2$(py)Cl and TiO$_2$ | Ru(dcb)$_2$(Cl)-bpa-Os(bpy)$_2$(Cl). Experiments on a large number of samples revealed some sample-to-sample deviation from which we conclude that the injection yield is the same to within 10%. The results demonstrate that Ru(II)* excited states in the model and bimetallic sensitizers lead to charge separated states with the same efficiency, but do not rule out a Ru(II)* injection mechanism where energy transfer to Os(II) is followed by Os(II)* injection for TiO$_2$ | Ru-bpa-Os. For this pathway to occur, intramolecular energy transfer would have to be kinetically competitive with electron injection. Considering known energy transfer rate constants for related Ru-Os polypyridyl compounds [10] and the ultrafast injection rates recently reported [12], this pathway is highly improbable.

Shown in Figure 5 are typical transient absorption responses assigned to interfacial charge recombination, Equation 4. The sample was excited with 532 nm light (8–10 ns fwhm, ~5 mJ/cm^2) in 0.1 M LiClO$_4$/CH$_3$CN under an argon atmosphere at 25°C. No significant differences were observed with 417 nm or 683 nm light excitation. The data could not be fit to a single first-order or a single second-order kinetic decay model.

No significant photocurrent is observed when TiO$_2$ | Ru-bpa-Os is used as a photoanode in a regenerative solar cell with 0.5 M LiI/ 0.05 M I$_2$ in acetonitrile. This is expected because the Os(III) center is a weak oxidant and charge recombination is faster than iodide oxidation [13]. The lack of a photocurrent indicates that iodide oxidation by Ru(III) formed after blue or green light excitation, does not compete kinetically with intramolecular electron transfer from Os(II).

CONCLUSIONS

In summary, the first bimetallic sensitizer designed to display diode-like behavior at semiconductor surfaces is reported. This interface clearly rectifies charge with forward electron transfer rates at least five orders of magnitude faster than charge recombination. We note that the design of this compound was largely inspired by the electropolymerized, spatially segregated, two-layer films of Ru(II) and Os(II) reported by Meyer and Murray several years ago [14]. In this work, a Ru(II) polymer layer insulates an outer Os(II) layer from the electrode and charge could be trapped as Os(III) for hours compared to the milliseconds for the molecular analog reported here. The flexible bpa ligand employed and the geometry about Ru(II) allow direct Os(II)* → semiconductor electron transfer and probably provides a direct pathway for charge recombination as well. Future studies will focus on control of the surface orientation and systematic tuning of the thermodynamic driving forces for the light driven electron transfer processes that control the optoelectronic properties of these fascinating molecular materials. It is interesting to note how research on dye-sensitized solar cells, carried out in many groups all over the world, contributes to new areas of science with applications in molecular electronics.

ACKNOWLEDGMENTS

I would like to thank my research advisor Dr. Gerald Meyer for his constant support and encouragement. Special appreciation must be given to Dr. Roberto Argazzi and Dr. Carlo A. Bignozzi from the University of Ferrara, Italy. The collaboration with this group has been extremely fruitful and enriching for me. Finally, I would like to thank the former and present members of the Meyer Group at Johns Hopkins: Al Abramson, Laura Bauer, Bryan Bergeron, Dr. Fereshteh Farzad, Paul Hoertz, Dr. Craig Kelly, Minh Ko, John O'Callaghan, Ping Qu, Mark Ruthkosky, Don Scaltrito, Dr. Jeremy Stipkala, Arnie Stux, Dr. Dave Thompson, Mei Yang, and Mark Zaros.

REFERENCES

1] V. Balzani and F. Scandola, *"Supramolecular Photochemistry"*; Ellis Harwood: Chichester, U.K. (1990).

[2] C.A. Bignozzi, J.R. Schoonover, and F. Scandola, "A supramolecular approach to light harvesting and sensitization of wide-bandgap semiconductors: Antenna effects and charge separation", *Prog. Inorg. Chem.* 44, 1–95 (1997).

[3] R. Amadelli, R. Argazzi, C.A. Bignozzi and F. Scandola, "Design of Antenna-Sensitizer Polynuclear Complexes. Sensitization of Titanium Dioxide with $[Ru(bpy)_2(CN)_2]_2Ru(bpy(COO)_2)_2^{2-}$", *J. Am. Chem. Soc.* 112, 7099–7103 (1990).

[4] B. O'Regan and M. Grätzel, "A Low-Cost, High-Efficiency Solar-Cell based on Dye-Sensitized Colloidal TiO_2 Films" *Nature* 353, 737–740 (1991).

[5] R. Argazzi, C.A. Bignozzi, T.A. Heimer, and G.J. Meyer, "Remote Interfacial Electron Transfer from Supramolecular Sensitizers" *Inorg. Chem.* 36, 2–3 (1997).

[6] a) R. Argazzi, C.A. Bignozzi, T.A. Heimer, F.N. Castellano, and G.J. Meyer, "Light-Induced Charge Separation across Ru(II)-Modified Nanocrystalline TiO_2 Interfaces with Phenothiazine Donors" *J. Phys. Chem. B* 101 (14), 2591–2597.(1997). b) R. Argazzi, C.A. Bignozzi, T.A. Heimer, F.N. Castellano, and G.J. Meyer, "Long-Lived Photoinduced Charge Separation Across Nanocrystalline TiO_2 Interfaces" *J. Am. Chem. Soc.* 117 (47), 11815–11816 (1995).

[7] P. Bonhôte, J.E. Moser, R. Humphry-Baker, N. Vlachopoulos, S.M. Zakeeruddin, L. Walder, and M. Grätzel, "Long-Lived Photoinduced Charge Separation and Redox-Type Photochromism on Mesoporous Oxide Films Sensitized by Molecular Dyads" *J. Am. Chem. Soc.* 121, 1324–1336 (1999).

[8] T.A. Heimer, S.T. D'Arcangelis, F. Farzad, J.M. Stipkala, and G.J. Meyer, "An Acetylacetonate-Based Semiconductor-Sensitizer Linkage" *Inorg. Chem.* 35 (18), 5319–5324 (1996).

[9] a) M.J. Powers, T.J. Meyer, "Medium and Distance Effects in Optical and Thermal Electron Transfer" *J. Am. Chem. Soc.* 102 (4), 1289–1297 (1980). b) S.A. Adeyemi, E.C. Johnson, F.J. Miller, and T.J. Meyer, "Preparation of Pyrazine-Bridged, Polymeric Complexes of Ruthenium(II)" *Inorg. Chem.* 12 2371–2374 (1973).

[10] B. Schlicke, P. Belser, L. DeCola, E. Sabbioni, and V. Balzani, "Photonic Wires of Nanometric Dimensions. Electronic Energy Transfer in Rigid Rodlike Ru(bpy)$_3$$^{2+}$-(ph)$_n$-Os(bpy)$_3$$^{2+}$ Compounds (ph = 1,4–Phenylene; n = 3, 5, 7)" *J. Am. Chem. Soc.* 121 (17), 4207–4214 (1999).

[11] C.A. Kelly, D.W. Thompson, F. Farzad, J.M. Stipkala, G.J. Meyer, "Cation-Controlled Interfacial Charge Injection in Sensitized Nanocrystalline TiO$_2$" *Langmuir* 15(20), 7047–7054 (1999).

[12] a) Y. Tachibana, J.E. Moser, M. Grätzel, D.R. Klug, and J.R. Durrant, "Subpicosecond Interfacial Charge Separation in Dye-Sensitized Nanocrystalline Titanium Dioxide Films" *J. Phys. Chem.* 100 (51), 20056–20062 (1996). b) T. Hannappel, B. Burfeindt, W. Storck, F. Willig, "Measurement of Ultrafast Photoinduced Electron Transfer from Chemically Anchored Ru-Dye Molecules into Empty Electronic States in a Colloidal Anatase TiO$_2$ Film" *J. Phys. Chem. B* 101 (35), 6799–6802 (1997). c) T.A. Heimer, and E.J. Heilweil, "Direct Time-Resolved Infrared Measurement of Electron Injection in Dye-Sensitized Titanium Dioxide Films" *J. Phys. Chem. B* 101 (51), 10990–10993 (1997). d) R.J. Ellingson, J.B. Asbury, S. Ferrere, H.N. Ghosh, J.R. Sprague, T. Lian, and A.J. Nozik, "Dynamics of Electron Injection in Nanocrystalline Titanium Dioxide Films Sensitized with [Ru(4,4'-dicarboxy-2,2'-bipyridine)$_2$(NCS)$_2$] by Infrared Transient Absorption" *J. Phys. Chem. B* 102 (34), 6455–6458 (1998).

[13] M. Alebbi, C.A. Bignozzi, T.A. Heimer, G.M. Hasselmann, and G.J. Meyer, "The Limiting Role of Iodide Oxidation in *cis*-Os(dcb)$_2$(CN)$_2$/TiO$_2$ Photoelectrochemical Cells" *J. Phys. Chem. B* 102, 7577–7581 (1998).

[14] H.D. Abruna, P. Denisevich, M. Umana, T.J. Meyer, and R.W. Murray, "Rectifying Interfaces using 2–Layer Films of Electrochemically Polymerized Vinylpyridine and Vinylbipyridine Complexes of Ruthenium and Iron on Electrodes" *J. Am. Chem. Soc.* 103 (1), 1–5 (1981).

PART II

SIMULATION-TRAINING

Out-of-Core Isosurface Visualization of Large Volume Data Sets

Peter D. Sulatycke

Department of Computer Science
State University of New York
Binghamton, NY 13902-6000
Research Advisor: Dr. Kanad Ghose

ABSTRACT

In earlier papers, interval trees were shown to optimally perform isosurface extraction of volume data by only accessing cells contributing to the isosurface. In this report we present a new in-core and out-of-core algorithm (span-space buckets) that also achieves this optimal property but with about half the storage requirement. Our technique thus retains the same level of performance as interval trees but can now handle larger data sets. We also introduce an interleaved version of out-of-core span-space buckets that additionally reduces seeking on the disk by a considerable amount, speeding up the overall isorendering time. Lastly, we reduce the out-of-core storage requirements of these algorithms by adapting an in-core chessboarding technique, introduced by Cignoni et al., to out-of-core use. This technique reduces out-of-core storage requirements by a further four fold without sacrificing performance. As a result, large data sets can be viewed on systems with limited RAM as fast as in-core isosurface extraction algorithms on systems without RAM constraints. Furthermore, the performance realized in viewing the visible woman data set on a PC is comparable to what has been realized recently on a 64–node supercomputer. Our in-core and out-of-core span-space bucket techniques are thus the technique of choice over techniques that use the interval tree and similar search-optimized data structures.

INTRODUCTION

Modern scientific applications generate three-dimensional data sets, also commonly referred to as volume data. Volume data can be generated directly from imaging systems (such as CT, MRI, PET and Ultrasound scans) or it can be "synthesized" through virtual environment creation (such as in medical and flight simulation training) and scientific simulations (such as weather prediction, airfoil analysis, stress analysis, climate modeling etc.) We consider the problem of visualizing isosurfaces within large volume data sets. Isosurface extraction for such volume data sets consists of three stages: (1) **locating intersecting cells:** the identification of cells that intersect the desired isosurface (i.e., cells that contain points where the isovalue occurs), (2) **surface modeling:** determining and modeling (typically with triangles) the exact manner in which the isosurface intersects these cells and (3) **display:** the actual display of the three-dimensional isosurface as a two-dimensional image on a computer screen. The focus of this report is on the evaluation of new techniques to accelerate Stage 1 of isosurface extraction of memory-resident and disk-resident rectilinear data sets, while also keeping storage requirement to a minimum. These technique can also be easily adapted to handle non-structured data and other structured data sets.

The last stage of isosurface extraction is usually performed by some type of graphics accelerator hardware. The second stage is typically performed by the well established marching cubes table lookup [1]. This algorithm uses a lookup table to determine the triangles that form the isosurface within an individual cell. The thrust of recent research has involved stage 1, searching for cells intersected by the isosurface [1-7]. The algorithm described in [8] is particularly effective because it obviates the need for cell searching, only cells that actually contribute to the isosurface are visited. This optimum property results from the use of an interval tree data structure [9].

The performance improvements provided by these cell extraction techniques usually come at a cost in the form of increased memory requirements to hold auxiliary structures. Most existing isosurface extraction algorithms generally assume the availability of enough fast local memory to store the complete data set and its auxiliary structures. To visualize any data set and auxiliary structures larger than the available RAM capacity, the user has to add more RAM or rely on virtual memory mechanisms to move data through the storage hierarchy. Specifically, when conventional visualization techniques are applied to visualize disk-resident volume data whose capacity exceeds that of the RAM, severe page thrashing occurs, causing the visualization time to increase unduly and well beyond real-time bounds. Even with the current availability of large amounts of memory, volume data can still exceed RAM capacity. For example, applying the algorithm presented in [8] to a 512x512x512 12–bit data set, would require over 1.25 Gigabytes of RAM. This problem is not going to improve with larger memory capacities because the data set sizes continue to grow as sampling and synthesis techniques improve.

The isosurface extraction technique presented in this report addresses this issue while maintaining optimum isosurface extraction capability. This technique, referred to as *span-space buckets*, provides virtually the same complexity of the interval tree but with a much lower overall storage requirement. This enables span-space buckets to nearly equal the fastest existing in-core isosurface extraction algorithms when sufficient RAM is available and to surpass existing performance when RAM is limited (relative to the data size). Even with the reduced storage requirement of span-space buckets, large volume data can still exceed RAM capacity. To address this issue, we also present in this report an out-of-core version of span-space buckets that is not affected by RAM limitations. We show that out-of-core span-space buckets nearly equals the performance of our recently introduced binary interval tree based out-of-core algorithm [11,12]. Additionally, we show that out-of-core span-space buckets equal or exceed the performance of in-core isosurface extraction algorithms. Lastly, we show how out-of-core interval trees and out-of-core span-space buckets can reduce their storage requirements by four fold without affecting performance. Before describing span-space buckets we first describe prior work that influenced this work.

PRIOR ISOSURFACE EXTRACTION ACCELERATION TECHNIQUES

Stage 1 of isosurface extraction simply entails finding the cells that are intersected by an isosurface defined by isovalue Q. A cell is intersected if the isovalue Q lies along the range (aka. interval) defined by the maximum and minimum data values that reside at the vertices defining the cell. For example, the range in rectilinear data is defined by the maximum and minimum of the eight data values that lie on the vertices of the cuboid. We refer to these values as the max value and the min value of the cuboid/cell. This problem is also referred to as the *stabbing point query* problem since the Q values stabs or intersects the interval. This problem can be formally stated as the following: given a set of intervals, $I = \{(a_1, b_1), (a_2, b_2), .. (a_m, b_m)\}$, where a_i and b_i are the two endpoints (i.e., extrema) of interval $I_i = (a_i, b_i)$, determine all the intervals that contain Q.

The notion of span-space aids greatly in the visualization and design of isosurface extraction [4]. The span-space, as shown in Figure 1 (a), represents intervals as points in a two dimensional space where the minimum values of intervals correspond to values on the X axis and the maximum

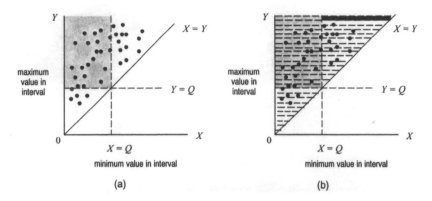

Figure 1. (a) span-space formulation of the isosurface extraction problem, (b) and its direct exploitation using the span-space bucket data structure.

value within an interval correspond to values along the Y axis. An interval (a_i, b_i) is represented as a point $(X = a_i, Y = b_i)$. Note that all of the intervals (each representing a cell) correspond to points above or on the diagonal line, $X = Y$. The intervals that intersect the isosurface for an isovalue of Q are points located in the shaded region shown in Figure 1 (a), bounded by the line $X = Q$ and $Y = Q$, as their max values (Y-axis) are higher than Q and their min values (X-axis) are lower than Q.

The problem of finding all cells within the shaded region in Figure 1 (a) is strictly a known computational geometry problem. One method of solving this problem is by sorting the cells by their max and min values to facilitate the detection of cells in the shaded region. In [4] the authors use a Kd-tree (a multi-dimensional variant of the binary search tree) to simultaneously sort all cells by max and min values. This algorithm, called NOISE (Near Optimal IsoSurface Extraction), has the desirable property of only requiring $O(N)$ storage and worse case query complexity $O(\sqrt{N} + K)$ (N is the total number of cells and K is the number of cells actually intersecting the isosurface). This complexity measure is not a true measure of the overall isorendering performance but a general guideline for comparison. For example, the time to perform a comparison of the cells in $\sqrt{N}$ is quite small compared to the time to produce triangles from the cells in K. In addition, such issues as caching, pointer indirections and book-keeping complexity are ignored. Actual comparisons of running times would be more meaningful but are hard to come by because source code is rarely freely available. Thus, worst case asymptotic bounds are usually given as an indication of performance.

In [8] Cignoni et al. proposed the use of binary interval trees to accelerate isosurface extraction. Binary interval trees were first proposed to handle the *stabbing point query* problem by Edelsbrunner in [9]. The worse case computational complexity of isosurface extraction was shown to be $O(h/2 + K)$, where h is equal to the number of bits used to represent the data values [8]. For example, if the data being visualized is stored in two byte integers, on average 8 cells will be compared that do not intersect the isosurface. The interval tree structure has the advantageous properties of requiring minimal book-keeping for traversal and nearly all data accesses are in-order within the interval tree. The in-order nature of interval trees not only aids processor caching when the interval tree is resident in RAM but it also allows out-of-core interval trees to easily be implemented [11,12]. The price that binary interval trees pay for this in-order access, is that each cell is required to be represented in the interval tree twice. This repeated storage is a natural consequence of the data being sorted by both minimum and maximum values. As will be seen, our span-space buckets algorithm maintains nearly the same computational complexity as interval trees but only requires the data to be stored once.

At about the same time Cignoni et al. proposed the use of interval trees to accelerate isosurface extraction [8], Shen et al proposed using a lattice subdivision of span-space for the same purposes [6]. This algorithm was referred to as ISSUE (Isosurfacing in Span Space with Utmost Efficiency). The stated worse case complexity of ISSUE was given as $O(\log (N/L) + \sqrt{N/L} + K)$ in [6], with L indicating the number of rows and columns used to divide span-space into a lattice. In [13], a variation of the ISSUE scheme is used to extract isosurfaces for time variant data. This paper indicates that the true worse complexity of ISSUE is $O(\sqrt{N/L} + K)$ because a binary search for cells within lattice regions is not needed. As in the case of interval trees, ISSUE has the valuable property of visiting cells in-order and the undesirable property of having to store cell information twice. Again this duplication of data is the result of sorting the data by maximum and minimum values.

As we will show in the next section the worse case complexity of our span-space bucket scheme is $O(K + 2^{h-1})$, (where h is still equivalent to the number of bits used to represent the data values). Unlike ISSUE and interval trees, span-space buckets does not require the data to be stored twice. In addition, it has minimal book-keeping, visits data in-order and is very conducive to out-of-core implementations.

SPAN-SPACE BUCKETS

Before describing span-space buckets in detail, we will describe in detail the ISSUE algorithm so that similarities and differences can be pointed out.

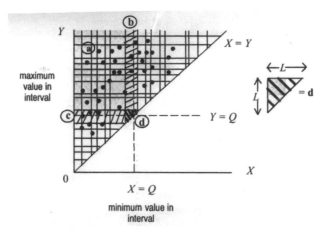

Figure 2. ISSUE algorithm: division of the span-space into L X L lattices.

Figure 2 depicts ISSUE's division of the span-space into L X L lattices. Each lattice cell effectively contains interval/cells within a L X L rectangular region within the span-space and is identified with a pair of indices. ISSUE's lattice subdivision of span-space, in conjunction with the $Y = Q$ horizontal and $X = Q$ vertical lines, produces 5 distinct regions within the span-space. This subdivision simplifies the extraction of cells bounded by the $Y = Q$ and $X = Q$ lines: all cells/ intervals within Region A are trivially accepted as cells that match the query, while all cells outside the marked regions are trivially rejected, in Region B and C cells must be located with a binary search and finally the cells in triangular Region D are located via a kd-tree or interval tree. To perform the binary searches, all cells within Region B must be sorted by their minimum value and all cells within Region C must be sorted by maximum value. Since this method must work for all isovalues, every cell will belong to Region B and C at some point. Thus the intervals/ cells within every lattice cell must store their data or equivalent pointer information in two lists; one sorted by the maximum values of the cells and another sorted by their minimum values. As noted in [13], there is no need for the binary searches in Region B and C because cell extraction can be done by starting at the max or min value within the list (depending on the region) and moving toward the Q value. This simplifies the extraction process but it does not eliminate storing the data twice. In addition the processing of cells lying in Region D is not trivial.

Our span-space formulation is based on the fact that input data values typically have a very small fixed data range. (We show later, how it most likely can be extended to handle larger data ranges.) For example, most CT data is limited to 12 bits of data which can only represent 4096 different data values. Given this limited range, span-space can be broken into one bucket for each possible data value; as depicted by the horizontal lines in Figure 1b. Each cell is placed in the bucket that matches the cell's maximum value. Within each bucket, we sort the cells in ascending order of their minimum value. The bucket-based sorting algorithm used to accomplish this sort is the inspiration for the name, *span-space buckets*. Each strip actually corresponds to a bucket in the sorting process. The use of a bucket for every value allows the data to be sorted by maximum and minimum values without having to store the data twice. The buckets sort the data by maximum value while the cells within the buckets are sorted by minimum value. Note that this can be reversed with the use of minimum buckets that contain cells sorted by their maximum value.

Our in-core implementation is very straightforward. Instead of storing cell data directly inside the buckets, we store indices to point to the actual cell data represented within the buckets. This helps to keep the memory requirement to a minimum. The cell data itself is spatially organized (i.e., ordered by their coordinate values and in the "raw" 3–d format). All buckets are stored consecutively in RAM. The number of cells represented in each bucket is maintained in an array and used to quickly jump to the beginning of each bucket. As intersecting cells are located, they are passed to a marching cubes like process to compute the surface passing through them as a set of triangles. Unlike other algorithms, no maximum or minimum values are stored within the bucket structures, saving considerable space. This does not affect performance because the marching cubes index produced in the process of generating triangles indicates if a cell is above or below a query value. The process of generating triangles uses the pointers within the buckets to directly access the raw 3–d data set. Gradients are computed on the fly to reduce RAM requirements.

Extracting Isosurfaces With Span-Space Buckets

Even though span-space buckets are discrete by nature, they can be used for extracting isosurfaces from discrete or non-discrete data sets with little modification. In either case, the query value Q can be either discrete or non-discrete. The extraction process is as follows:

Discrete Data Values: Processing starts with the bucket for the highest possible maximum value. Cells within this bucket, starting with the smallest minimum value until a cell with a minimum value of $\lfloor Q \rfloor$ or larger is encountered, are sent to a surface modeling process based on the marching cubes lookup table. This process then continues to the next bucket and so on, ending with and including the bucket corresponding to the value of $\lceil Q \rceil$. This effectively processes all cells with a maximum value > Q and a minimum value ≤ Q.

Span-space buckets thus process all cells that an isosurface intersects and in the worst case only visits one bad cell in each bucket visited. This gives the previously described average complexity of $O(K + 2^{h-1})$ for isosurface extraction.

Non-Discrete Data Values: When the input data consists of non-discrete data, the span-space buckets creation and isosurface extraction process must be modified slightly. In this case a bucket for each discrete value is still maintained but now cells are placed in a bucket that corresponds to the floor of a cell's maximum value. If the Q value is also discrete, (i.e. the isosurface occurs at a discrete isovalue) query processing uses the above described method for discrete data values. If the Q value happens to be non-discrete than the process is slightly modified. Within a bucket, cells are now processed until a cell with a minimum value of Q or larger is encountered, not $\lfloor Q \rfloor$ as in the discrete case. The extraction process then continues to the next bucket and so on, ending with and including the bucket corresponding to the value of $\lfloor Q \rfloor$, not $\lceil Q \rceil$ as in the discrete case. Within the last bucket, a small number of cells will have a maximum value less than Q and will be not be processed.

The pros and cons of the span-space bucket data structure are as follows. As in the case of the previously described algorithms, span-space buckets is optimal in nature. Unlike ISSUE and interval trees, the data for a cell (or pointers to data) is not duplicated and it is in-order (exploiting efficient caching). Additionally, the creation of the span-space buckets can be quickly done in 2 or 3 passes of the data by the use of counting sorts. The only con of this algorithm is that the number of buckets that can be created is limited. We can easily create buckets for up to 20 bit data. After that size, several consecutive data values will have to be mapped to the same bucket to allow the scheme to be used with little modification. Such collapsing of a large data range should not be a problem since data values will be more sparsely located when the data range goes up. In addition, we have found in practice that up to half of all buckets are empty and do not need to be maintained. Note, that this mapping will only cause the last bucket processed to have non-intersecting cells within it. Thus we expect that span-space buckets will be efficient with any type of data, although this has not been confirmed as of yet.

OUT-OF-CORE IMPLEMENTATIONS OF SPAN-SPACE BUCKETS

Even with span-space buckets' reduced memory requirements, the algorithm will eventually degrade as the size of the data and the data structure exceed the available RAM capacity. Thrashing will occur and out-of-core techniques must be used. Out-of-core techniques avoid performance degradation due to thrashing, but to achieve reasonable performance out-of-core visualization techniques must minimize the amount of time for I/O processing. Span-space buckets is a good structure for minimizing I/O since it optimally retrieves the cells that intersect an isosurface. In addition to keeping the amount of data read to a minimum, optimizing seek time is equally important in keeping I/O time down. This is especially true of modern disks because seek times have not been improving at the rate at which transfer rates have been improving: disk seek times have remained fairly flat in the range of 6.5 msecs. to 9 msecs., while transfer rates have gone up due to the widening of transfer paths and the steady increase in rotational speed from 5,400 r.p.m. to 10,000 r.p.m. To keep seek time down, out-of-core implementations of span-space buckets can not use pointer indirections to locate cell data as in the in-core case. Thus all data associated with a cell (including gradients) must be explictly stored within the span-space bucket structure. This not only minimizes the amount of seeking but it keeps all disk accesses in consecutive order. On the downside, this causes the volume data to be duplicated within several cells. Since this structure is out-of-core, this duplication is easily tolerated. Note, that this duplication is necessary in any out-of-core technique that stores data in line to avoid disk seeking. Lastly, we overlap I/O processing with the triangulation of cells by using threads in conjunction with mutex protected swinging buffers. This effectively hides most of the I/O processing time. Further details of this threads implementation can be found in our out-of-core interval tree papers [11,12].

For out-of-core implementations of the span-space buckets, a *naive implementation* will be to directly use the in-core implementation, as shown in Figure 1 (b), with data for each bucket laid out on the disk from the top to the bottom. In other words, the data for the bucket of cells with a max value of *max_value* is stored on the disk, followed by the data for the bucket of cells that have a max value of (*max_value* −1) and so on, ending with the data for the bucket of cells that have a max value equal to *min_value*. Within each bucket, the data is stored in increasing order of the min value of the cells. The top diagram in Figure 3 (a) shows the layout of the first four buckets when written to disk. Each bucket contains cells with the same max value and ascending min values, while each block within each bucket represents cells with the same min value. Cell extraction proceeds in the same manner as used in the in-core implementation, i.e.. reading buckets one at a time and in-order. Even though buckets are read in-order, significant seeking can occur since a trailing region of each bucket must be skipped. In span-space the trailing section for the first bucket is shown in dark grey in Figure 1 (b). While the bottom diagram in Figure 3 (a) depicts the disk skipping that must occur due to these trailing sections. We have observed that the seeking needed to skip over this data negatively affects I/O performance even with 8 bit data (255 buckets).

Reducing I/O Time With Interleaved Span-space Buckets

One way of reducing the seeking necessary to retrieve all isosurface intersecting cells is to interleave the buckets when they are written to disk. What results is called *interleaved span-space buckets*. The top diagram in Figure 3 (b) shows the disk layout of four buckets when their data is interleaved with each other. As in the previous diagram, each block within each bucket represents data with the same min value. Thus reading cells from the smallest min value to larger min values can be done in one sequential read. The bottom diagram in Figure 3 (b) depicts the sequential reading of cells contained in the first four blocks of all four buckets. This greatly reduces the seeking required for this particular query but it can worsen the seeking of other queries. For example, if only cells from the first three buckets need to be read in Figure 3 (b), every fourth block of cells (checkered) will have to be skipped or unnecessarily read in. In fact, if all buckets are interleaved the only thing that has really been done is to change the buckets from horizontal max buckets to vertical min buckets. Thus the overall amount of seeking hasn't changed. This can be seen in Figure 4 (a) which

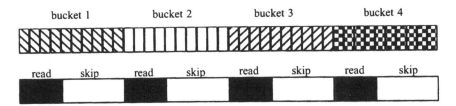

Figure 3. (a) Naive disk layout of the first 4 span-space buckets, (each block represents cells at the same minimum value). Followed by the reading and head skipping that takes place when reading the first 4 blocks in each bucket.

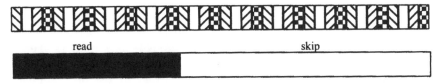

Figure 3. (b) The disk layout of the first 4 buckets using interleaved span-space buckets (buckets are now interleaved). Followed by the reading and head skipping that takes place when reading the first 4 blocks in each bucket.

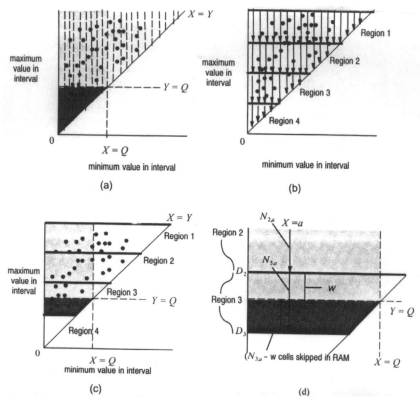

Figure 4. (a) Vertical min buckets in span-space; the darkly shaded area indicates cells that must b skipped over when out-of-core. (b) Vertical min buckets broken into four separate regions. Regions are written to disk in numerical order, with min buckets being written in left to right order and with cell within the bucket being written in the direction of the arrow. (c) With multiple regions the darkly shaded area that must be skipped is reduced. (d) The darkly shaded area is so small that it is read into memory and skipped there. A count w and length N are used to perform the skip.

depicts vertical min buckets in span-space. Instead of having to skip over the unshaded triangular area, as with horizontal max buckets, the darkly shaded triangular region must now be skipped.

A solution to this is to break span-space into regions and interleave the buckets within each individual region. Figure 4 (b) shows span-space divided into four such regions. Each region is written to disk one after each other, i.e. Region 1 followed by Region 2 and so on. Within each region the horizontal max bucket data is interleaved when written to disk. Figure 4 (b) visually shows this interleaving with arrows indicating the ordering of the data as it is written to disk. Each arrow represents cells at the same min value within a region and thus each arrow is actually a bucket at a particular min value. These buckets are written to disk in a left to right ordering (smallest min value to largest). Note, this is *not* at all like the lattice method of [13], buckets are still being used. You can think of the regions as being max buckets containing vertical min buckets. This is somewhere between using only horizontal max buckets or only vertical min buckets. As will be seen in the results section, significant reduction in total seek time occurs when span-space buckets are interleaved in this fashion.

The reason for this reduction can be seen if we step though the isosurface processing of Q in Figure 4 (c). The lightly shaded area in Region 1 can be read without any seeking and the unshaded area in Region 1 can be skipped with one seek operation. The total amount of data skipped over hasn't changed but the number of seeks has been reduced, saving a significant amount of time. Processing for Region 2 has the same behavior as Region 1. Processing of subsequent regions continues in this fashion until a region is intersected by the Y = Q line. For the example shown, Region 3 is intersected and must be processed differently. In this region, either the darkly shaded section can be skipped over on disk or it can be read into RAM when each of the vertical min buckets are processed. If this data is read into RAM then the unwanted cells can easily be skipped in memory. The extra time required to read in the extra data is insignificant as long as the regions are fairly small, thus this is the preferred approach. On the downside, using small regions means more regions are needed and the time saved can be overshadowed by the extra seeking required for more regions. Thus regions should be small but not too small. Before describing the technique used to determine region sizes, we more formally describe the isosurface extraction processing steps.

As in the in-core case, out-of-core span-space buckets can be used for discrete and real data values. For brevity sake, the following description will be limited to the implementation of span-space buckets on the disk when the cell data values are discrete, ranging from *min_value* upwards to *max_value*. With slight modifications, similar to the ones described in the in-core query processing section, non-discrete data values can also be handled. Note, even though the data is discrete the following technique works for discrete and non-discrete isovalue queries. To read in all intersecting cells for any given isovalue, the following pieces of information are needed: the position of the horizontal lines that demarcate the different regions, the length of each vertical min bucket in each region and the length of each region. This information naturally comes out of the creation process of the interleaved span-space buckets and is stored on disk until processing begins. It is then brought into memory and stays there until the user is finished with all isosurface extraction. The lengths of the regions do not always have to be stored on disk since they can be calculated from the lengths of the min buckets. We will use the notation to designate the horizontal demarcation line of region x, which is denoted as (note the value is the smallest max value in). We will also denote N_{xy} as the length of min bucket y in region x.

Out-of-Core Processing Steps:

1. Seek to the beginning of the data on disk (i.e. the beginning of R_1)
2. If the region R_x is above or intersected by the horizontal line (Y = Q) for the given isovalue Q, (i.e., the value of D_x is $\geq \lceil Q \rceil$ or $D_{x-1} > \lceil Q \rceil$), read all cells within R_x that have a min value less than Q (i.e. cells left of the X = Q vertical line) as follows:

— compute the sum of min bucket lengths for all min buckets less than Q:
$$S_x = \sum_{y=0}^{y=\lfloor Q \rfloor -1} N_{xy}$$
— read off the data for cells from the disk into the RAM (these are in consecutive locations, so no disk locations are skipped)
— seek to the beginning of the next region (R_{x+1}) by using the length of R_x and then go to the beginning of step 2.

In-Core Processing Steps:

3. The processing of in-core cells are done in the same order as they were read in, (i.e. starting with the first region and ending with the intersected region, from min bucket 0 to min bucket $(\lceil Q \rceil -1)$ and the largest max value to smallest within each min bucket). Since all the cells read into RAM from regions above or just on the horizontal line (Y = Q) must be intersected by the isosurface, compute the isosurface going through these cells. This is done by modeling the surface with triangles produced through the use of a marching cubes like lookup table. For the one region that is intersected by the (Y = Q) line, some of the cells read into RAM are not intersected and must be processed as follows:

— A counter, denoted as w, is used to keep track of the number of cells that actually intersect the isosurface. At the start of processing of each min bucket the counter is initialized to zero. Since we are processing the cells in the same order as they were read in, all isosurface intersecting cells within a bucket are processed in sequential order. Thus when it is determined that a cell does not intersect the isosurface, we can conclude that all of the remaining cells within the same min bucket are also non-intersecting. The check for intersection is an implicit part of the marching cubes algorithm used to compute the surface, so the check for the first non-matching cell does not constitute any extra overhead. On finding the first non-intersecting cell, $(N_{xy} - w)$ cells are skipped in RAM to get to the start of the next min bucket. See Figure 4 (d) for an example of the cells that are skipped for the min bucket at (X = A). (Cells in the darkly shaded area are skipped in RAM). Processing of the remaining min buckets from this one region are done in the same manner. As mentioned previously, the number of non-intersecting cells read into memory is kept to a minimum by using many small regions.

Determining the Size and Number of Regions

Choosing the exact size of each of the regions is complicated by the fact that the amount of disk seeking should not depend heavily on the query value, irrespective of the data distribution for the cells. To simplify the region sizing problem, we first make the assumption that the distribution in span-space is uniform and relax this assumption later. At first glance it might appear that making each region contain the same number of cells would produce consistent reduction in seeking across all isovalues but this is not the case because span-space is triangular. The regions produced by this span-space division will cause isosurface extraction of higher isovalues to perform better, at the cost of lower valued isovalues. A division of span-space that prevents isosurface extraction from favoring any isovalue can be obtained if a uniform lattice is superimposed on span-space, such that each lattice cell contains the same number of cells/intervals. The positions of the horizontal lattice lines are used as the demarcation lines (D_x) of each region, while the vertical lines are not needed. On relaxing the uniform distribution constraint, an effective division can still be attained if the lattice cells still maintain the same number of cells. To do this, lattice columns and rows are non-uniformly spaced. Note, the lattice is only used to figure out the region's demarcation lines (D_x), buckets are still used throughout the algorithm. Instead of actually applying a lattice to span-space the following indirect method is used:

1. A predetermined number of regions, say N, are chosen and these regions are numbered from top to bottom, starting with 1. (Note this is opposite of the region ordering as depicted in the figures.) N is dependent on the size of the complete out-of-core span-space bucket structure and the number of bits used to represent the data. For a 6.8 Gigabyte span-space bucket structure, with 12 bit data, 200–400 regions are needed. More data sets need to be tested before a heuristic can be developed.
2. Assuming Z to be the total number of cells in the data, the demarcation between consecutive regions are chosen to ensure that region k (for k ranging from 1 to N) roughly contains $Z * (k/(N*(N + 1)/2)$ cells. This

equation arises from the fact that a lattice superimposed on span-space will cause each region R_x to have one more lattice cell than region R_{x+1}. The position of the horizontal lines that demarcate the regions are recorded as D_x for later use.

REDUCING DISK STORAGE NEEDS WITH CHESSBOARDING

When used for structured data, the storage requirement of out-of-core span-space buckets is large compared to the size of the raw data set. The process of transforming the raw data set into the span-space bucket structure considerably destroys the spatial coherence in the data. As a consequence, the cell's coordinates and sometimes gradients have to be stored explicitly within the out-of-core span-space bucket structure. However, since the data for a cell has to be stored only once within the out-of-core span-space bucket, the overall storage requirements is about half that of other value based isosurface extraction algorithms. (These other algorithms require the cell data, coordinates, and possibly gradients, to be written to disk twice.) As introduced in [8], chessboarding can be used to reduce the memory requirements for value based isosurface extraction of structured data sets. We adapt this in-core technique to reduce storage requirements for out-of-core isosurface extraction algorithms.

The basic idea behind chessboarding is to recognize that the edges and vertices of cells are shared with neighboring cells. This sharing of edges implies that if a cell is intersected by an isosurface than the adjacent cells that share the cell's intersected edges must also be intersected by the isosurface. Thus with the proper organization, information doesn't need to be explicitly stored for every cell. Chessboarding provides this type of organization by classifying cells into black and white cells. Each cell consists of eight data points at the eight corners of a cubic cell (or a rhomboid cell, for non-rectilinear data spaces). Information, such as pointers to the cells for in-core span-space buckets, only need to be stored for the black cells. As will be seen, all information about white cells can be gathered from the neighboring black cells in *all three* dimensions.

If the cell space is viewed as slices, as shown in Figure 5, consecutive *interior* slices have black and white cells arranged as shown. Figure 5 does not shown the extremum layers of the chessboarded data sets. For convenience, they can be considered as being composed of black cells entirely—the resulting data on the extremum slices are somewhat redundant, but this redundancy can be tolerated. The basic chessboarding scheme, as described above, has roughly 3 times as many white cells as there are black cells, so that the data compaction achieved is roughly a factor of 4. Finally, note that although

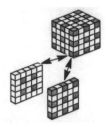

Figure 5. A chessboarded volume data set; the slices at the extremities along each three dimension have been taken off. The first two slices of the interior slices of this chessboarded data set are shown on the left. The information for the data points within a white cell are implicit in the neighboring black cells in its own layer and the two surrounding layers. The extreme layers that have been "peeled" off for this diagram need special considerations (see text).

the term chessboard has been used to allude to the layout of black and white cells, the resulting pattern does not strictly resemble a chessboard, where the squares in adjacent layers alternate between black and white.

This chessboarding layout was used in [8] to reduce the pointer storage requirement of in-core interval trees. As presented by Cignoni et al. in [8] and detailed in [7], the associated modeling technique did not fully exploit the information known about neighboring white cells nor was the resulting isosurface extraction algorithm appropriate for an out-of-core implementation. Our chessboarding technique is grossly different from the original scheme presented in [8], so much so that it is a fundamentally different technique, albeit with the same name. It is thus useful to first discuss the chessboarding scheme of [8] in detail to bring out these differences.

The implementation of in-core interval trees by Cignoni et al [8], uses an interval tree data structure, with pointers from the left and right lists pointing to the cell data within the 3–dimensional raw data set (see [9] or [11] for the basic concepts relating to an interval tree). This avoids the inclusion of any cell data within a list, the lists only contain pointers to a common data set. The storage savings realized through the use of pointers are seriously offset by the storage requirements of the pointers themselves. To reduce the storage impact of the pointers, Cignoni et al used the chessboarded scheme to store pointers to only the black cells within the interval tree. The black cells, accessible through the the pointers, were then used as seeds to visit neighboring white cells that also intersect the isosurface of interest. This is easily accomplished due to the fact that when a white cell is intersected by an isosurface, its intersected edges are also the intersected edges of neighboring black cells. The data needed to compute the isosurface for the white cells is accessed directly from the raw data set. Since multiple black cells, acting as seeds, can point to the same white cell, repeated computations of the surface going through that white cell were avoided by marking the white cell as visited. In summary, Cignoni et al. used chessboarding to reduce pointer storage without exploiting the full potential of chessboarding.

Out-of-Core Chessboarding Implementation

To avoid random seeking on disk, out-of-core span-space buckets can not contain pointers to access an out-of-core raw data set. Instead the data within each cell must be stored explicitly within the span-space bucket structure. Since the implicit coordinate information of the raw data set is not available, coordinate information for each cell, along with information on all eight vertices must be directly included in the data structure. This results in a huge amount of redundancy that can be avoided if chessboarding can be adapted to out-of-core use. Fundamentally it is easy to see that if all the triangles are known for the black cells then the all the triangle vertices of the white cells must also be known. What is not immediately know is which vertices go with which white cells and how are they organized to form triangles. Note, this is a different problem then just using black cells to locate white cells as in the in-core case. If this was in-core, one could look at the vertex values that make up the intersecting cell to determine an index into a marching cubes like lookup table. We will show that by just storing a little information on each white cell after each black cell calculation, the white triangles can be quickly created once all black cells are processed. The overall approach is as follows:

Do Steps 1 and 2 for each black cell extracted from the disk-resident tree into the memory (RAM):

1. **Calculate Black Cell Triangles:** As black cells are being read in from memory, calculate the triangles that model the isosurface going through the black cell. After each cell is processed, store the resulting coordinate and gradient information for each triangle vertex in the next available space in an array (Figure 6).
2. **Propagate Black Cell Information to Adjacent White Intersecting Cells:** First, determine which white cells share an intersecting edge with the black cell. Second, pass on the sign information (greater than or less than the isovalue) of each of the vertices from the black cell's intersecting edges to the white cells that share these edges. Third, for each of these white cells store pointers to the appropriate black cell triangle vertices stored in 1.
3. **Calculate White Cell Triangles:** Once all black cells are calculated use the information stored in step 2 to quickly form the white triangles.

The structures used to store the information in step 2 are depicted in Figure 6, along with the array used in step 1. Since white cells may be adja-

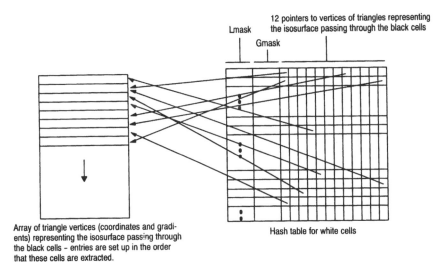

Array of triangle vertices (coordinates and gradients) representing the isosurface passing through the black cells - entries are set up in the order that these cells are extracted.

Hash table for white cells

Figure 6. The hash table and array structures used to compute the isosurface going through the white cells.

cent to multiple black cells intersected by the isosurface, information stored on white cells must be repeatedly accessed. To accomplish this we use a hash table for all white cells that are indexed with the coordinate value of the white cell. Within each white cell's hash entry, two 8–bit bit masks that contain the known sign information (greater than or less than the isovalue) for all eight cell vertices are stored. The first bit mask (gmask) contains ones for each vertex value known to be greater than the input isovalue (threshold). The second bit mask (lmask) contains ones for each vertex value known to be less than the input isovalue. We will soon describe how these two masks can be used to quickly calculate the marching cubes lookup index in step 3. In addition to storing the bit masks, pointers are stored to the black triangle vertices (stored in the array in step 1) that lie on the shared interesting edges. This data will later be used to form the white triangles.

Implementation Details of Black Cell Propagation

Since Step 2 is performed multiple times on each white cell, it must be made very efficient. To that end we use a series of lookup tables to propagate black cell information to the white cells:

- First, we must quickly determine for a given black cell which white cells share an an intersected edge with the black cell. Normally this information is determined as each black intersecting edge is visited but we can not afford to do that because it would mean multiple hashing for each white cell or the white cell location will have to be temporarily stored. We get around this by storing a small lookup table that indicates exactly which white cells share an interesting edge with the black cell. The index created for the marching cubes table lookup is used to index into this table.

- Second, we use another lookup table to quickly propagate information to the lmask and gmask of the white cells. For each unique marching cubes case, we predetermine which white cell vertices must be greater than or less than the isovalue. This is done by storing for each black cell case in the lookup table, a less than and greater than bit mask for each white cell that shares an intersecting edge with the black cell. For each white cell, the bit masks from the lookup table are AND-ed with the lmask and gmask stored with a white cell's hash table entry. This effectively updates the white cell's bit mask with new information about its vertices.

- Lastly, a third lookup table is used to determine which black triangle vertices go with which white cells' edge. Again this information is all precalculated and stored in a lookup table that is indexed by the black cell's marching cube index.

Details of White Cell Triangles Calculation—Fast Sign Propagation

Once all black cells are processed, the stored bit masks are used to calculate the triangles in the white cells. For all white cell vertices that make up an intersected edge the two bit masks indicates which side of the surface the vertex lies. This "inside-outside" information is not immediately know for vertices that do not lie on an intersected edge. This information must be calculated. Once the inside-outside information is known for each vertex the marching cube table index is likewise known. This index and the pointers to the calculated black cell intersecting vertices and gradients are used to write out all of the white cell's triangles; no gradient and intersection calculations need to be done, as these were computed and saved at the time when a black cell was processed.

Determining the inside-outside information for vertices on non-intersected edges can easily be done by propagating sign information from known verti-

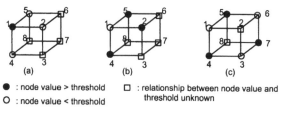

● : node value > threshold   □ : relationship between node value and
○ : node value < threshold   threshold unknown

Figure 7. Three example scenarios in handling the white cells.

ces. Although this is easy to do it is not very efficient. What follows is a quick way of calculating the unknown vertices from the lmask and gmask. First the number of bits turned on is calculated for both the masks, gmask and lmask (this is quickly done with a small lookup table). If the two counts add up to eight then the marching cube table index is trivially equal to the gmask. Likewise, if the lmask count is greater that the gmask count, the gmask is trivially accepted as the marching cube table index. This can be seen in Figure 7 (a) where vertex 1 is greater than and three vertices are less than. In this case all unknown vertices must be less than the isovalue because there is a less than vertex between all unknown vertices and the intersecting surface. Similarly, when the the gmask count is greater than the lmask count, the unknown vertices must be all be greater than. This can be seen in Figure 7 (b). The bits representing all unknown vertices are turned on in the gmask to form the marching cube index. The last possible gmask and lmask combination is when the number of ones in both are equal to three. Such a situation is depicted in Figure 7 (c). The missing vertex inside-outside information can be determined by visiting a neighbor cell or by using a lookup table with the exact precalculated answer. There are only 8 possible cases where the counts will both be 3. By using the last 6 bits of the gmask each of these cases can be distinguished by looking into a table.

Lastly, the previous description assumed that only one span-space bucket structure existed for the raw volume data set. In reality this is not a viable solution because the number of white cells stored in the hash table will be too large. Thus multiple span-space bucket structures, encompassing several slices, are preferred with out-of-core chessboarding. To accommodate multiple structures, a circular set of three white cell hash tables and black cell triangle buffers are used. Note, one also has to be careful not to use too few slices per structure because this will cause excessive seeking of data on disk. We found that 6 to 10 slices typically produces good results. We found the performance of isosurface rendering with our chessboarding scheme to be very close to that of our out-of-core interval trees and span-space buckets without chessboarding. This is because of the compensation of the added computational effort needed for chessboarding by the need to only process about one-fourth of the cells and by the correspondingly reduced disk I/O needs. Thus, substantial reduction of the data set is achieved without any appreciable drop in performance. In addition, chessboarding is independent of span-space buckets and may be used with any value based out-of-core isosurface extraction structures.

Differences Between In-core and Out-of-Core Chessboarding

In spite of the similar goal of reducing storage requirements, both Cignoni et al.'s algorithm [8] and our out-of-core chessboarding technique are fundamentally different. First, our out-of-core chessboarding technique allows white cell triangles to be calculated without any direct access to the white cells. This is very unlike the in-core method that has direct access to the vertices of the white cells from the raw data set. Given this requirement, Cignoni's in-core chessboarding is not viable as an out-of-core technique. Second, unlike the approach taken in [8], we do not directly compute the isosurface going through the white cells. Instead as we process a black cell we store the the intersection points of the isosurface with the black cell edges (which are shared with the neighboring white cells) within data structures that are associated with the neighboring white cells. Finally when all the black cells are processed we have all the intersection points of the isosurface with all of the white cells. In [8], white cell triangles are calculated just like black triangles; chessboarding just helps to find the white cells. Each vertex of the white cell must be visited to determine the marching cubes index value.

EXPERIMENTAL RESULTS AND DISCUSSIONS

We evaluated the performance of both in-core and out-of-core span-space buckets with 8–bit CT data of the human head, supplied by the University of North Carolina (256 X 256 X 225) as well as the visible woman data set [10]. The human head data set was 14.7 Mbytes in raw form and took 65.6 seconds to be converted to an out-of-core span space bucket taking up 379 Mbytes. The host was a 450 MHz, 256 Mbyte Pentium II PC running Linux, employing Ultra DMA EIDE drives and a high-end ES Lightning 1200 graphics card. All results are the harmonic average of 5 runs, with each isovalue being tested consecutively.

Figures 8 through 10 depict the results for the human head data set. Specifically, figures 8 and 9 indicate how the in-core and out-of-core renderers using span-space buckets compare against some aggressive in-core and out-of-core visualizers. The total time plotted in the graphs refers to the time needed to produce the triangles for the surface to be rendered into a RAM buffer. This time does not include the time needed to actually display the triangles, which is mostly a function of the capabilities of the graphics card. For very large data sets the graphics card tends to become the bottleneck but with ongoing advances in graphics card design this is quickly changing. Additionally, techniques can be used to decimate the number of triangles in order to limit the (effective) required triangle display rate to match the performance of current graphics cards. The number of extracted intersecting cells varies from 85 cells for isovalue 255 to 1,550,143 for isovalue 9 and 1,179,388 for isovalue 89.

The results shown in Figures 8 and 9 demonstrate that both the in-core interval tree and the in-core span-space buckets provide comparable performance, 1.5 to 2.5 times faster than the *optimized patented* marching cubes

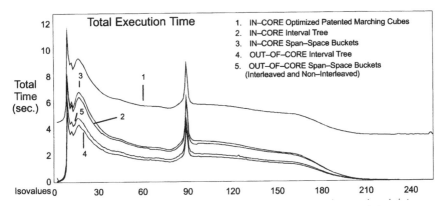

Figure 8. Total time need for generating triangles for the human head data.

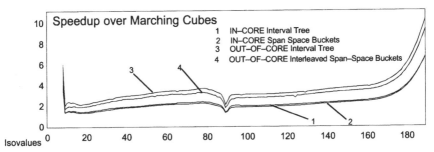

Figure 9. Speedup over the optimized & patented marching cubes technique for the human head data.

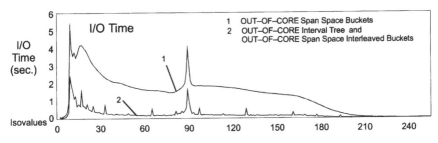

Figure 10. I/O times for the out-of-core techniques for the human head data.

technique over a wide range of isovalues. (The optimized version of the patented matching cubes algorithm includes several new optimizations, particularly for caching that improves the performance of the patented version by as much as 30% or so.) We were not able to make comparisons with other isosurface extraction algorithms because of their limited public availability. The marching cubes algorithm is depicted in the figures because it provides a good baseline for comparison. The out-of-core versions of the span-space buckets, with and without interleaving, provides a performance level that is 2 to 4 times better than that of their in-core counterparts. Its performance also approaches quite close to that of our out-of-core interval trees [12], which is also plotted in these graphs. The out-of-core algorithms provide better performance because of two reasons: first, gradients are already calculated for the out-of-core algorithms while the in-core algorithms calculate gradients on-the-fly and second, the out-of-core algorithms better exploit processor caching since all of there RAM access are in-order. (Note, to reduce seeking gradients had to be precomputed for the out-of-core algorithms while the in-core algorithms had to calculate gradients on the fly because of limited RAM, otherwise thrashing would occur.) In the speedup graphs of Figure 9, the speedup of all of the techniques go up considerably compared to the patented, optimized marching cubes technique. This is due to the fact that for the isovalues beyond 180, the number of cells intersecting the isosurface is very small compared to the total number of cells in the data. (The plots of Figure 9 confirm this statement.) The interval tree and span-space bucket based algorithms do considerably well here as they only visit the cells that intersect the isosurface: a manifestation of the optimal nature of these techniques. In contrast, the marching cubes technique visit all of the data cells, whether they intersect the isosurface or not. We also expect our out-of-core span-space interleaved bucket based algorithm to perform better than the out-of-core techniques of [14,15], both of which do not address the problem of seeking on the disk.

The performance edge of the out-of-core span-space interleaved buckets comes about because of the dramatic manner in which it reduces the disk I/O time through the use of an organization that reduces disk seek times significantly. The I/O times for the span-space interleaved buckets and that for the out-of-core interval tree are virtually identical, as shown in Figure 10. However, the span-space interleaved buckets occupy about half as much space on the disk as the interval tree data structure for the same raw data set. Figure 10 demonstrates how effectively the interleaved organization of the out-of-core span-space buckets reduces disk I/O time over the naive implementation of span-space buckets on the disk. Much of this reduction comes from a drastic reduction in the course of extracting cells from the disk into the RAM. Note that the I/O times plotted in Figure 10 are *overlapped with the computation times* because of the use of multithreaded implementation. The relatively small I/O times are effectively hidden from the computations of the triangles, leading to the low isorendering times shown in Figure 8 for all of the out-of-core techniques. Interleaved span-space buckets does not outperform span-space buckets in this case because all of the I/O time for both cases is completely hidden.

Interleaving of span-space buckets becomes more important as the data range of the data set goes up and seeking requirements increase. This can be seen in our next data set, the first 512 slices of the visible woman data set [10]. This data is 512 X 512 X 512, 12 bit data, that requires 256 MBytes of storage. Since 12 bit data requires 4096 different buckets, more seeking is involved with this data set. This data set took 603 seconds to be prepro- cessed into an out-of-core span-space bucket requiring 6.9 GBytes of stor- age. Nearly all of this preprocessing time results from I/O time. Due to a 2G file limit, the span-space bucket structure was written to several files. Ap- proximately 5 GBytes of this data is devoted to gradients; this could easily be reduced by reducing the number of bits used to store the gradients. As noted previously, the large size of the out-of-core data is due to the explicit incorpo- ration of data into each cell stored, necessary to reduce disk seeking.

Figures 11 and 12 show the performance of the out-of-core algorithms and the in-core patented marching cubes. Note, in-core interval trees is not practical with this size of data because of thrashing and thus is not dis- played in the graphs. Out-of-core interleaved span-space buckets and out- of-core interval trees perform approximately 1.5 to 10 times faster than marching cubes, with a 3.5 speedup at isovalue 600. Speed up for isovalues

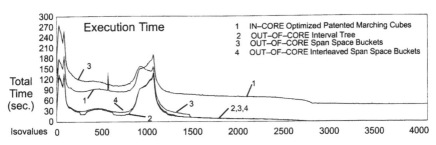

Figure 11. Total time need for generating triangles for the visible woman's torso.

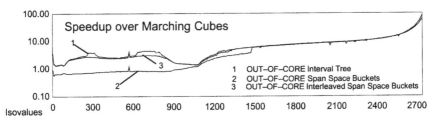

Figure 12. Speedup over the optimized & patented marching cubes technique for the visible woman's torso.

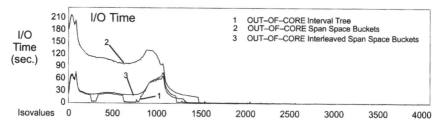

Figure 13. I/O times for the out-of-core techniques for the visible woman's torso.

over 2700 are not shown because they are so great. As in the previous data set this is because of the limited number of cells for these isovalues. The number of cells varies from 8 cells for isovalue 4090 to more than 18M cells for isovalue 82 and 1074. Note the poor performance of non-interleaved span-space buckets for isovalues less than 1000. This is caused from seeking and increased I/O time, as seen in Figure 13. By interleaving span-space buckets, I/O time drops down to the level of interval trees and total execution time decreases in response. Examples of images rendered using the out-of-core span-space interleaved buckets are depicted in Figure 15. The skin image was taken at isovalue 600 and took 31 seconds to produce the triangles. The bone image was taken at isovalue 1300 and took 18.7 seconds. This level of performance is very favorable compared to previously published work. For example, the technique in [16], takes approximately 42 seconds on a 64 processor Cray T3e to render isovalue 117 for the top torso of the visible woman (by matching number of triangles we believe that the first 512 slices were used for this result), while we take 54 seconds to render the same isovalue on a 450Mhz PC (using interleaved span-space buckets.

Lastly, we conducted tests using our out-of-core chessboarding with both out-of-core interval trees and span-space buckets. In both cases the disk storage requirement was reduced by four fold, with a resulting decrease in overall I/O time. Additionally, the overall performance of both chessboarded algorithms only varied within 5% of their non-chessboarded counterparts. Due to limited space we only present the results for our out-of-core chessboarded interval trees on the complete visible woman data set [10], as seen in Figure 14. This data set is much larger than the available RAM (256M) of our testbed; consisting of 909 Megabytes of 12 bit data (512 X 512 X 1734). In-core techniques thrashed heavily on this data set and are not shown in the figure. The results presented in Figure 14 again show how effectively I/O and triangulation computations are overlapped through the use of multithreading. Examples of images of the complete visible woman are depicted in Figure 15; the skin image took 59.52 seconds to produce the triangles while the bone image took 29.82 seconds. The disk requirements

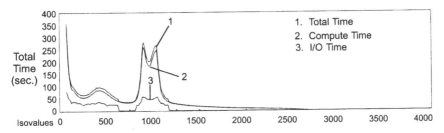

Figure 14. Out-of-Core chessboarded interval trees: total time for the complete visible woman data set (1734 slices).

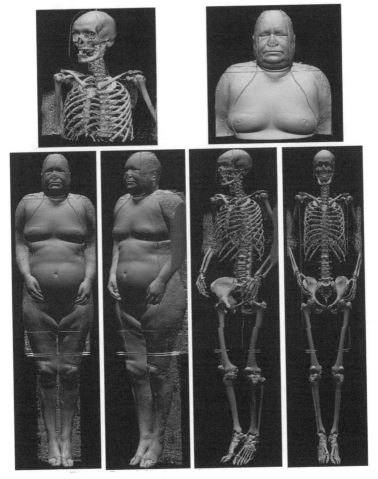

Figure 15. Example images from the Visible Woman 12 bit data set. Upper torso 512x512x512, Full body 512x512x1734.

of the chessboarded out-of-core interval trees for this data set were approximately 12 Gigabytes. With chessboarded span-space buckets this was reduced to 6 Gigabytes with little reduction in performance; making this the method of choice. This implies that for a given upper bound on the available storage, one can extract isosurfaces from larger raw data sets using chessboarded span-space buckets than one would using interval trees and at the same time get a performance very close to that of interval trees.

CONCLUSIONS

The span-space buckets technique introduced in this paper represents an inherently simplistic and intuitive way of performing isosurface extraction—both in-core and out-of-core. Our approach was inspired by the span-space concept introduced in [4]. The span-space bucket approach is search optimal—only the cells that intersect an isosurface are visited during processing. The span-space bucket structure is well-suited for out-of-core implementations but has slower I/O processing times than out-of-core interval trees. With the use of interleaving, the amount of seeking on the disk is greatly reduced, leading to an overall performance that matches that of the very optimal interval trees. In general, the storage requirements of our out-of-core span-space bucket techniques are approximately half the size of out-of-core interval trees and possible out-of-core implementations of IS-SUE [6]. We also introduced an out-of-core version of chessboarding that reduces the storage requirements of any value based isosurface extraction, including span-space buckets and interval trees, by a further four fold. In any case, the increasing sizes of modern disks to 40 GBytes and beyond, suggests that storage requirements will not be an impediment if fast isosurface extraction performance is required for large data sets. Currently the main bottleneck remains displaying the produced triangles in a timely fashion. Modern graphics card vendors are working on promising solutions in this respect. With the advent of new graphics cards, our out-of-core interleaved span-space bucket technique promises to allow very large data sets, such as the visible man and visible woman, to be viewed on low cost PCs at a reasonable level of performance: much faster than what can be realized using fast, traditional in-core techniques. Our basic technique can thus allow larger data sets to be viewed on PCs cost-effectively and in a more scalable manner than in-core techniques. Additionally, its performance is very close to that of the interval tree based algorithm. This makes the span-space bucket technique a viable approach for a wide range of volume data.

REFERENCES

[1] W.E. Lorensen and H.E. Cline, "Marching Cubes: A High Resolution 3D Surface Construction Algorithm, Computer Graphics," vol. 21, no. 4, pp.163–169 (1987).
[2] R.S. Gallagher, "Span Filter: An Optimization Scheme for Volume Visualization of Large Finite Element Models," *Proc. Visualization '91* Conf. Proc., pp 68–75 (1991).
[3] T. Itoh and K. Koyyamada, "Isosurface Generation by Using Extrema Graphs," *Proc. Visualization '94.* Los Alamitos, Calif.: IEEE Press, pp. 77–83 (1994).
[4] Y. Livant, H. Shen, and C. Johnson, "A Near Optimal Isosurface Extraction Algorithm for Structured and Unstructured Grids," *IEEE Trans. Visualization and Computer Graphics,* vol.2, no. 1, pp. 73–84, Apr. 1996
[5] H. Shen and C.R. Johnson, "Sweeping Simplices: A Fast Iso-Surface Extraction Algorithm for Unstructured Grids," *Proc. Visualization '95,* pp. 143–150 (1995).
[6] H. Shen, C.D. Hansen, Y. Livnat, and C.R. Johnson, "Isosurfacing in Span Space with utmost Efficiency (ISSUE)," *Visualization '96 Conf. proc.,* pp. 287–294 (1996).
[7] P. Cignoni, personal communication, 1998.
[8] P. Cignoni, P. Marino, C. Montani, E. Puppo, R. Scopigno, "Speeding Up Isosurface Extraction Using Interval Trees", *Visualization and Computer Graphics,* vol. 3, no. 2, pp. 158–170 (1997).
[9] H. Edelsbrunner, "Dynamic Data Structures for Orthogonal Intersection Queries," Technical Report F59, Inst. Informationsverarb., Tech. univ. Graz, Graz, Austria (1980).
[10] The Visible Human Project http://www.nlm.nih.gov/research/visible/visible_human.html
[11] P. Sulatycke and K. Ghose, "Out-of-Core Interval Trees for Fast Isosurface Extraction," in Proc. Late Breaking Topics, *IEEE VIsualization '98 Conference,* pp. 25–28. (Available at: http://opal.cs.binghamton.edu/~sulat.) (1998).
[12] P. Sulatycke and K. Ghose, "A Fast Multi-threaded Out-of-Core Visualization Technique," in *Proc. IEEE 13–th Intl. Symposium on Parallel and Distributed Systems and the 10–th Symposium on Parallel and Distributed Processing (merged symposia),* pp. 569–575 (1999).
[13] H. Shen, "Isosurface Extraction in Time Varying Fields Using a Temporal Hierarchical Index Tree", in *Proc. IEEE Visualization '98,* pp. 15–166 (1998).
[14] Y. Chiang, C. Silva, "I/O Optimal Isosurface Extraction", *Proceeding of Visualization '97,* pp. 293–300, Phoenix AZ, Oct (1997).
[15] Y. Chiang, C. Silva, "Isosurface Extraction in Large Scientific Visualization Applications Using the I/O-filter Technique", unpublished paper, from Y. Chiang (1997).
[16] L. Bajaj V. Pascucci D. Thompson X. Y. Zhang, "Parallel Accelerated Isocontouring for Out-of-Core Visualization", *Proceedings of IEEE Visualization '99, Parallel Visualization and Graphics Symposium,* pp. 97–104 (1999).

Exploiting Temporal Uncertainty in Simulation Models Using a Cluster Computing Environment

Margaret L. Loper

Georgia Institute of Technology

College of Computing

Atlantic Drive

Atlanta, GA 30332–0280

Research Advisor: Dr. Richard Fujimoto

ABSTRACT

Time management is required in simulations to ensure the simulation
model correctly reproduces temporal aspects of the system under investigation. One might think that you just "hook together the simulators" and
have them send messages to each other. However, this can lead to problems in that the simulated world may not correctly reproduce temporal
aspects of the real world that is being modeled. The importance of correctly reproducing temporal relationships depends on the simulation application. In analytic simulations impossible event orderings could cause a
simulator to fail. In training simulations non-causal event orderings may
not be perceptible to human participants if they occur in rapid succession.
Since analytic and training simulations have different time management
requirements, they are not used together in federations. However, one can
envision reuse of analytic simulations in training environments to populate a virtual environment intended for human-in-the-loop tank simulations. The key to achieving this interoperability is a time management approach that relaxes the ordering constraints of the analytic simulations in
order to achieve faster execution.

A new time management mechanism, called Approximate Time, has been
proposed to solve this problem. Approximate Time exploits temporal uncertainty in the simulation by using time intervals rather than precise time
values. Approximate Time allows the modeler to carefully tailor ordering
constraints according to model and usage requirements in order to facilitate simulation reuse across domains. This fellowship investigated two
issues related to Approximate Time: what size time intervals can be used
without invalidating the simulation model and how to assign timestamps
from the interval in a consistent way across the federation.

PROJECT DESCRIPTION

Time management is required in simulations to ensure the simulation model correctly reproduces temporal aspects of the system under investigation. One might think that you just "hook together the simulators" and have them send messages to each other. However, this can lead to problems in that the simulated world may not correctly reproduce temporal aspects of the real world that is being modeled. For example, delays in the real world depend on quantities such as the speed of light whereas delays in the simulated world depend upon quantities having nothing to do with the simulation model, e.g., network delays. This can lead to anomalies such as the cause appearing to happen *after* the effect. The importance of correctly reproducing temporal relationships depends on the simulation application. In analytic simulations impossible event orderings could cause a simulator to fail. On the other hand, in training simulations non-causal event orderings may not be perceptible to human participants if they occur in rapid succession.

Time management in the High Level Architecture (HLA) [1] is concerned with the mechanisms for controlling the advancement of time during the execution of a federation. A problem not addressed by HLA time management mechanisms concerns the reuse of simulations across traditionally different application domains. For example, one can envision reuse of event driven simulations in training environments to populate a virtual environment intended for human-in-the-loop tank simulations. However, synchronization overhead may prevent the event driven simulations from achieving real-time performance. Therefore, one would like to relax the ordering constraints of the event driven simulations in order to achieve faster execution. A mechanism that allows the modeler to carefully tailor ordering constraints according to model and usage requirements would facilitate reuse across domains.

There is one approach to time management that exploits temporal uncertainty in the simulation model. Specifically, two new message ordering services, called Approximate Time (AT) and Approximate Time Causal (ATC) order, which use simulation time intervals rather than precise time values, have been proposed to replace time stamp order event processing [2]. Each event X is assigned a time interval $[E(X), L(X)]$ where $E(X) < L(X)$, and $E(X)$ denotes the earliest point in simulation time that the event may occur and $L(X)$ the latest. Time intervals are assigned to the event by the simulation model, typically based on uncertainty regarding when the event occurs. For example:

- The time a moving vehicle comes into range of a sensor depends on factors such as the speed the vehicle is traveling and the sensitivity of

the sensor, which in turn depend on environmental conditions and other factors that cannot be known with complete certainty.
- The think time required by human operators in issuing orders in a military simulation cannot be determined with complete certainty.
- Job service times such as the amount of time for a bank teller to serve a customer cannot be known with complete certainty.

It is clear that temporal uncertainty is ubiquitous in simulation modeling, stemming from the fact that simulations are only an approximation of the real world. The goal of AT and ATC order is to exploit this uncertainty to enhance the performance of parallel simulations.

Initial experiments indicate this approach shows promise in achieving good performance and the gains realized had negligible impact on the numerical results produced by the simulator. However, the results described in [2] scratch the surface in the use of AT and ATC ordering for parallel and distributed simulations. Numerous issues remain to be investigated. Two issues were investigated:

- *Model Validation.* What size time intervals can be used without invalidating the simulation model?
- *Timestamp Assignment.* In simulations where there are multiple recipients of messages for the same event, how are timestamps assigned from the interval in a consistent way across the federation?

MODEL VALIDATION

An issue remaining for AT-and ATC-order is the maximum time interval size that can be used without invalidating the simulation model. Since the size of time interval defines the uncertainty that can be tolerated for the event, the larger the time interval the less certain the result for that event. For a specific event this may have little significance. However, over the course of a simulation execution, the uncertainty could add up to cause events to happen in different orders. While this may be a valid outcome for some simulations, for others it could invalidate the results of the execution.

For example, Figure 1 shows a federation that is composed of two blue tanks (B1 and B2), a blue platoon leader (BPL), two red tanks (R3 and R4), and red platoon leader (RPL). The behavior of the tanks to the platoon leader is as follows. When a tank detects an enemy, it sends a "Detect" event to its platoon leader. The platoon leader then sends the tank a "Start Fire" (or SF) event indicating it is ok to fire at the enemy. Once the tank receives a SF event, it will send a "Fire" event followed by an "Update"

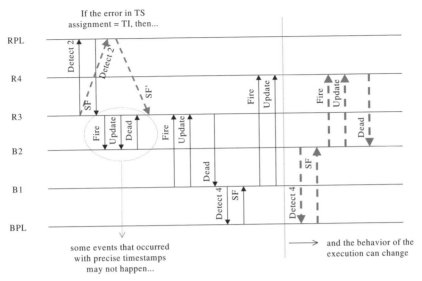

Figure 1. Example of how large uncertainty can change the outcome of an execution

event, updating the state of its attributes. If the Fire event caused the enemy tank to be killed, the enemy tank will send a "Dead" event in response to the Fire.

In the example, the "normal" execution is shown in solid lines:

Red Tank 3 detects, fires and kills Blue Tank 2
Blue Tank 1 detects, fires and kills Red Tank 3
Blue Tank 1 detects and fires at Red Tank 4

By introducing a large interval size for the Detect and Start Fire events, we can see how the uncertainty can change the outcome of the execution. The "modified" execution is shown in dashed lines:

Red Tank 3 detects Blue Tank 2
Blue Tank 1 detects, fires and kills Red Tank 3
Blue Tank 1 detects and fires at Red Tank 4
Blue Tank 2 detects, fires and kills Red Tank 4

In the new execution, we see that, due to the uncertainty of the time when the Detect and Start Fire events occurred, Blue Tank 2 is not killed by Red Tank 3. Instead Blue Tank 2 goes on to Red Tank 4. One might initially think this error completely invalidates the results of the execution (as compared to the execution using precise time stamps). And in fact, this may be true for some models. However, in others this new sequence of events may be perfectly valid. In fact, the uncertainty associated with the Detect 2 and Start Fire 2 events maybe be an accurate reflection of a communication problem the Red Tanks are having with their Platoon Leader. There-

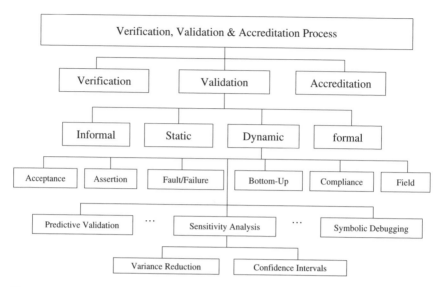

Figure 2. Approach for Determining Interval Size.

fore, the uncertainty the time interval introduce to an event is a better reflection of the interaction being modeled.

This research addresses the issue of how large a time interval can be before invalidating the model. An approach was developed (shown in Figure 2) for selecting the interval size, which is part of the Verification, Validation, and Accreditation (VV&A) process. Since the amount of uncertainty associated with the interval could invalidate the simulation execution, we believe that determining the maximum interval size is part of *validating* the simulation. Since the maximum size of the time interval is best determined by evaluating the execution behavior (as shown in the example above), a *dynamic technique* is required to validate the model execution. Our approach uses a *sensitivity analysis* by systematically changing the size of the time interval and observing the effect upon model behavior. The size of the time interval can also be changed to induce errors and determine the sensitivity of model behavior to such errors. In other words, the sensitivity analysis can identify the interval sizes to which model behavior is very sensitive.

The remainder of this section is organized as follows. First, an overview of the VV&A process is given including the basics of Validation and VV&A in the M&S life cycle. This section is to give the reader the necessary background on VV&A if they do not already have it. Next, dynamic verification and validation techniques will be discussed including sensitivity analysis. Finally confidence intervals are described in the context of validating time interval size.

A Methodology for Selecting Time Interval Size

According to [3], *"Verification* ensures that a simulation meets all the requirements specified by the user and that it implements those requirements correctly in software; *validation* ensures that a simulation conforms to a specified level of accuracy when its outputs are compared to some aspect of the real world." As shown in Figure 2, we believe that the time interval size should be determined as part of the validation process. Since the size of the interval indicates the amount of uncertainty the model can handle, the greater the uncertainty (i.e., large the interval) the less accurate the model may become.

Background on Validation

According to DoDD 5000.59, validation is "the process of determining the degree to which a model is an accurate representation of the real world *from the perspective of the intended uses of the model."* In other words, there must be a clear understanding of the intended uses of the model and a clear definition of the real world. Knowledge of how the simulation will be used determines the degree of detail that must be represented for the simulation to provide usable results. It also determines the degree of correspondence with real-world phenomena that will be sufficient to use the simulation with confidence.

A simple form of validation consists of comparing the output of a simulation with an observation from the real world, and then making a decision whether the result is good enough for your problem. Another validation technique includes testing simulation performance against a range of conditions, and then comparing the results with other models and simulations known to have validity in that operating range or comparing the results with the opinion of subject matter experts (SME). This latter technique is called sensitivity analysis.

As described in [3], validation is typically addressed at two levels. *Conceptual Model Validation* is the determination (usually by a group of SMEs) that the assumptions underlying the model are correct and that the proposed simulation design (i.e., the simulation's functions, their interactions, and outputs) likely will lead to results realistic enough to meet the requirements of the application. *Results validation* compares the responses of the simulation with known or expected behavior from the subject it represents in order to determine that responses are acceptable for the range of intended uses of the simulation. This process includes comparing simulation outputs with the results of controlled tests, sensitivity analyses, or expert opinion.

Determining the maximum time interval size falls into the *Results validation* category since we are trying to determine if the results of the execution using a time interval size greater than zero is sufficiently accurate for the range of intended uses of the simulation. This will be accomplished by a

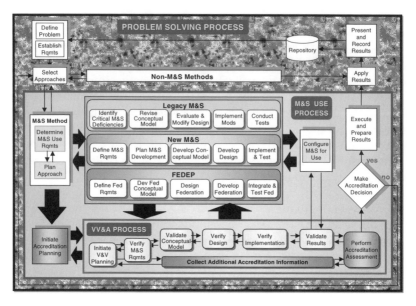

Figure 3. VV&A process in the M&S life cycle (taken from [3]).

sensitivity analysis, which varies the time interval size over a range of values $(0 \rightarrow \text{max\_boundary})$ and then compares the output with the expected behavior of the system.

The VV&A Process and Selecting Time Interval Size

The VV&A process is an integral part of the M&S life cycle, as shown in Figure 3. Since the primary purpose of the VV&A effort is to establish the credibility of the model or simulation, selecting the time interval size is part of this process. As shown across the bottom of Figure 3, the VV&A process has six primary steps: initiate V&V planning, verify M&S requirements, validate conceptual model, verify design, verify implementation, and validate results. Our methodology is integrated into the VV&A process at three of these steps, as described below.

Initiate VV&A Planning

The first step in the VV&A process is to develop a plan that identifies the tasks, requirements, resources and timelines required to VV&A the application. Formal guidance and requirements are collected and reviewed to determine the constraints under which the VV&A efforts will operate. The plan includes appropriate evaluation techniques and measures, as well as tools, resources and the schedule of specific activities. Initially, the plan is developed as a draft or working document that evolves as the application takes shape. Tailoring, the selection of verification and validation techniques based on requirements and resource availability, is done as part of the VV&A planning process to determine the most appropriate and cost-effective ways to address the application requirements and acceptability criteria.

It is in this phase that the V&V techniques (sensitivity tests) are defined and designed for time intervals. The first step is to define the accepted model behavior (part of Conceptual Model Validation) which includes fidelity, accuracy, and credibility requirements. Next, the simulation developer must identify those events (object updates and interactions) that will use time intervals. Because a simulation can use time intervals and precise time stamps for different events during an execution, the developer must decide which events can use time intervals. The last step is to design the sensitivity tests. This includes determining the test conditions, the valid time interval ranges (for each type of event), and the time interval ranges that will be used to induce errors. In addition to developing the test plan, any tools necessary to collect and analyze the time interval data must be identified at this point.

V&V the Application

Once the model or simulation is ready to be run, the application context needs to be verified and validated, as shown in the last two steps of the VV&A process. This includes such housekeeping tasks as ensuring that the appropriate platforms are being used and that operators and humans-in-the-loop are properly trained.

It is at this stage that the time interval sensitivity tests are performed. The experiments are conducted and data is collected according to the VV&A plan developed. The simulation model behavior is evaluated by analyzing the data collected. The output of the tests may reveal unexpected effects, which could indicate invalidity in model. In the analysis, any errors identified may help determine the sensitivity of model behavior to time interval size. Identifying time interval sizes to which model behavior is sensitive is a goal; model validity can be enhanced by ensuring time intervals for those messages are specified with sufficient accuracy. Ultimately, the results of the sensitivity analysis will be evaluated by comparing the results of the simulation execution with known or expected behavior from the subject it represents (e.g., controlled tests or expert opinion).

Verification and Validation Techniques

In the previous section, we described how selecting time interval size could be integrated into the VV&A process. The advantage of this methodology is that it leverages off a well-known process that is part of the M&S life cycle. This section will discuss in more detail the techniques used in that process to select the interval size.

Verification and Validation

Informal	Static	Dynamic	Formal
Audit	Cause-Effect Graphing	Acceptance Testing	Induction
Desk Checking	Control Analysis	Alpha Testing	Inference
Face Validation	Calling Structure	Assertion Checking	Logical Deduction
Inspections	Concurrent Process	Beta Testing	Inductive Assertions
Reviews	Control Flow	Bottom-Up Testing	Lambda Calculus
Turing Test	State Transition	Comparison Testing	Predicate Calculus
Walkthroughs	Data Analysis	Compliance Testing	Predicate Transformation
	Data Dependency	Authorization	Proof of Correctness
	Data Flow	Performance	
	Fault/Failure Analysis	Security	
	Interface Analysis	Standards	
	Model Interface	Debugging	
	User Interface	Execution Testing	
	Semantic Analysis	Monitoring	
	Structural Analysis	Profiling	
	Symbolic Evaluation	Tracing	
	Syntax Analysis	Fault/Failure Insertion Testing	
	Traceability Assessment	Field Testing	
		Functional (Black-Box) Testing	
		Graphical Comparisons	
		Interface Testing	
		Data	
		Model	
		User	
		Object-Flow Testing	
		Partition Testing	
		Predictive Validation	
		Product Testing	
		Regression Testing	
		Sensitivity Analysis	
		Special Input Testing	
		Boundary Value	
		Equivalence Partitioning	
		Extreme Input	
		Invalid Input	
		Real-Time Input	
		Self-Driven Input	
		Stress	
		Trace-Driven Input	
		Statistical Techniques	
		Structural (White-Box)	
		Branch	
		Condition	
		Data Flow	
		Loop	
		Path	
		Statement	
		Submodel/Module Testing	
		Symbolic Debugging	
		Top-Down Testing	
		Visualization/Animation	

Figure 4. Taxonomy of V&V Techniques (taken from [3]).

A taxonomy of V&V techniques is shown in Figure 4. According to [3], V&V techniques fall into four categories: informal, static, dynamic, and formal. The use of mathematical and logical formalism in each category increases from informal to formal, from left to right. The complexity also increases, as the category becomes more formal.

Dynamic V&V

As stated previously, our methodology uses a Dynamic V&V technique to select time interval size for a simulation. Dynamic V&V techniques evaluate the model based on its execution behavior. Most dynamic V&V techniques require model instrumentation or the insertion of additional code into the model to collect information about the model's behavior during execution. The type of instrumentation is determined manually or automatically based on static analysis of the model's structure.

Dynamic V&V techniques usually are applied in three steps: (1) the executable model is instrumented, (2) the instrumented model is executed and the model output is analyzed and (3) dynamic model behavior is evaluated.

To illustrate the dynamic V&V technique, consider the GTank federation described in [4]. GTank is a military federation composed of two types of federates: tanks and platoon leaders. The federates can be instrumented in Step 1 to record the following information: (a) time a tank detects an enemy vehicle; (b) time a tank fires at an enemy vehicle; (c) tank's position when enemy was detected; and (d) time that a tank was killed by enemy fire. In Step 2, the federate is executed and the information collected is written to an output file. In Step 3, the output file is examined to reveal discrepancies and inaccuracies in model representation.

Sensitivity Analysis

Sensitivity analysis is a dynamic V&V technique performed by systematically changing the values of model input variables and parameters over some range of interest and observing the effect upon model behavior [5]. Sensitivity analysis is used to determine if the simulation output changes significantly when the value of an input parameter is changed, when an input probability distribution is changed, or when the level of detail for a subsystem is changed [6]. Sensitivity analysis can identify those input variables and parameters to which model behavior is very sensitive. Model validity then can be enhanced by ensuring that those values are specified with sufficient accuracy [7] [8] [9].

When performing sensitivity analysis, it is important to use the method of common random numbers to control randomness in the simulation. Otherwise, the effect of changing one aspect of the model may be confounded with other changes that inadvertently occur.

To illustrate sensitivity analysis, again consider the GTank federation [4]. The federates were each instrumented to record the information identified in the last section. For each run, the same input parameters and random number seeds were used in order to keep the federation's initial state constant. The federation was executed multiple times, each time changing only the time interval size associated with object updates and interactions. In the federation, all federates used the same interval size and all events

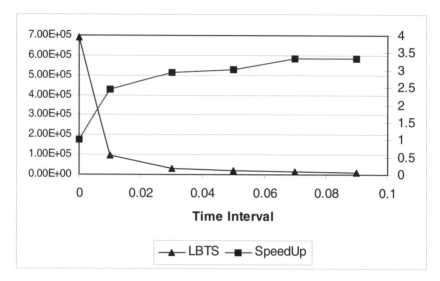

Figure 5. Speedup and Number of LBTS Computations for AT Federation.

(objects updates and interactions) used time intervals instead of precise time stamps. The federation was executed and the results of the federation were collected and analyzed. For each run, the sensitivity of "time of death" was analyzed to determine how the size of the time interval affected the battle outcome. Those results are given in [4] and summarized below.

The difference in the "time of death" of a tank using time intervals as compared to the time stamp order (TSO) delivery is shown in Figure 5. As would be expected, the larger the time interval the larger the error in time of death. This is because the error in assigning the time stamp can be as large as the time interval, since the time interval defines the error bound that can be tolerated for that event.

As seen in Figure 6, for time intervals less than or equal to .05 the error is less than 1 second. However, for time intervals greater than .05, the behavior of the simulation changes. For example, with a time interval less than .05, the error in time of death of blue tank 3 is small, meaning it was killed within 1 second of when it was killed when time intervals were not used (i.e., using the traditional TSO delivery mechanism). However, when the time interval was greater than .05, the outcome of the tank engagement was different enough that blue tank 3 ended up not being killed at the same time. In fact, when the time interval was .07, the difference in time of death was almost 90 seconds different! This indicates that the engagement blue tank 3 was part of was very sensitive to time. By introducing a large temporal uncertainty (>.05) in event times, it was possible to change the

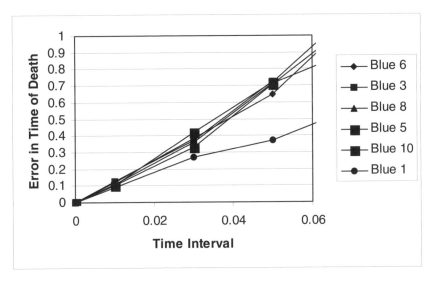

Figure 6. Difference in Time of Death Using Time Interval.

time at which blue tank 3 was detected and fired upon by enemy tanks. This, therefore, changed the outcome of the battle.

Confidence Intervals

A confidence interval is a notion from probability. It is commonly used to determine if sets of data are consistent with previous or present sets of data. A Confidence interval is a range of values that span from the Lower Confidence Limit to the Upper Confidence Limit. We expect this range to encompass the population parameter of interest, such as the population mean, with a degree of certainty, which we specify up front. The degree of accuracy we expect from the determined confidence interval is $1-\alpha$, where we pick α to be an acceptable risk of being wrong. For instance, we are willing to take a 5% chance of being wrong (α), so we expect that the confidence interval which we calculate will have a 95% chance of actually containing the population mean value between the lower and upper bounds, or confidence limits.

We can use confidence intervals to compare two systems on the basis of some performance measure (e.g., time of death of tank [4]). The comparison of the performance measure forms a confidence interval for the difference in the two systems. The confidence interval will quantify how much the performance measure differs, if at all, between the two systems. This idea can be illustrated in the GTank federation by comparing the "time of death of tank X" performance measure from the federation that uses precise time steps (time interval = 0) with the federation that uses a time intervals (time interval > 0).

Summary

This section has described a methodology for selecting time interval size in an Approximate Time Causal order federation. The methodology is based on Validation of a simulation model and can be easily integrated into a simulation model's VV&A process. We have described the steps in the VV&A process that must address time intervals and have described the specific V&V techniques to be used in the methodology. We also illustrated how the V&V techniques would be used with the GTank federation described in [4].

TIME STAMP ASSIGNMENT

This section is concerned with the timestamp assigned to an event (message) when it is sent to multiple destinations. Specifically, if an event is sent to multiple federates, it may be important that the same precise time stamp be assigned to the event across the different federates receiving the message. An efficient mechanism is needed to enforce this property. This problem arises in the current implementation of the AT-order mechanism because time stamps are assigned at the *receiver* prior to delivery of the message. Unless some additional mechanism is put into place different receivers may assign different time stamps to messages corresponding to the same event.

A related question concerns the federate sending the original message. In the current HLA definition, federates do not receive notification (e.g., via Reflect callbacks) of any message it sends. Thus, if a federate generates a message with a time stamp interval, how does it know what time stamp is eventually assigned to that event? This is in part an HLA Interface specification [1] issue; though it does impact to some extent how the federate is written. From a HLA Run Time Infrastructure (RTI) perspective, an important question is how does one ensure the precise time stamp is reported back to the federate sending the message, and how does one ensure this time stamp is the same as that assigned by the other federates receiving the message. This is effectively an instance of the "multi-destination problem" described above, i.e., any solution to the multi-destination problem should also be applicable to solving this problem. In the following section, both of these issues are addressed.

Solution Approaches to the Multi-Destination Problem

Three methods for ensuring messages sent to multiple destinations (which may include the federate sending the message) for a single event all receive the same assigned time stamp are described below. These solutions include:

1. *Receiver-Independent assignment:* The basic idea is to define a function to select the assigned time stamp that will compute the same value independent of the processor on which it is executed. If this function is given the same inputs on each of the processors receiving the message, it can be used to compute the same assigned time stamp on each of these processors, as well as the processor sending the message. The challenge with this approach is the time stamp assignment is constrained by values computed by the synchronization algorithm. In general, these constraints will differ from one processor to the next, invalidating the "receiver independent" requirement. However, it happens that for certain AT/ATC synchronization algorithms, these constraints are in fact the same for all processors, enabling a simple approach to achieve consistent assignment of time stamps. In particular, the ATC synchronization algorithm currently in place provides a very simple solution to this problem, as elaborated upon below.
2. *Time stamp interval assignment:* precise time stamps (rather than intervals) can be used for those events where consistent time stamp assignments for multiple destination messages are needed. Clearly this diminishes the benefit of using intervals, so this approach is desirable if only a limited number of events require consistent time stamps in the federation.
3. *Pre-sampling:* This approach uses a conventional RTI that delivers messages to each federate in time stamp order. Federates still utilize time intervals. Specifically, the *sender* computes the assigned time stamp (for the interval) by drawing a random number, places this time stamp in the message, and sends it as a TSO message. Of course, this approach is doomed to fail if the application has poor lookahead. To address this problem, the random number generator is sampled *in advance,* prior to when the time stamps are needed, and these pre-sampled values are used to improve the lookahead of the simulation. The advantages of this approach are that it allows time stamp intervals to be exploited on a conventional RTI, and that it offers the modeler greater statistical control over the assigned time stamps. Though at first glance this appears to offer a viable approach, closer inspection reveals that like most lookahead exploitation approaches, this approach is relatively "fragile" and requires some relatively strong assumptions on the federate to yield a correct realization.

The remainder of this section is organized as follows. First, issues concerning notifying the sender of the assigned time stamp are discussed. This is followed by a discussion of two of the approaches for ensuring consistent assignment of time stamps for multiple destination messages. Specifically, the first approach appears the most promising of the three, and is described. The second is fairly straightforward, so will not be discussed further here. Finally, the third approach is described in greater detail to highlight the restrictions that must be put into place to produce a correct implementation. Performance results are also included for this approach.

Sender Notification

An issue of concern is that a federate sending a message with a time interval may need to know the final time stamp assigned to the event. This may be important because the federate generating the event may also model one or more entities for which the event is relevant. The current HLA I/F Spec provides no provision to provide information back to the federate concerning a message it had previously sent. For example, attribute updates sent by a federate are not reflected back to the sender, even if the sender is subscribed to receive the information it produced.

Ways in which the I/F Spec might be modified to address this issue are discussed next. The first approach is applicable to the receiver-independent assignment function and time stamp interval assignment approaches to solving the multiple destination problem. The second is appropriate for the pre-sampling approach.

1. An obvious solution is to reflect messages produced by a federate back to that federate if it specifies this is to occur. For example, a parameter could be added to the Update Attribute Values and Send Interaction services (and others sending time managed messages) to indicate the federate is to receive Reflect Attribute Values / Receive Interaction callbacks for the message once the time stamp assignment has been made. Alternatively, a separate callback could be used with the assigned time stamp and a handle for the event (the Retraction handle could be used) to pass the assigned time stamp back to the federate.
2. If the assigned time stamp is known when the message is sent, this value could be returned to the federate in the service sending the message.

With either solution, an inconsistency could occur if a federate passes an event both to the RTI, and directly to other entities within the same federate. This is because a precise time stamp is used for the events passed directly between entities within the federate, while a time interval is used for events sent between federates. The time stamp selected from the interval may not match that used for events passed between entities within the same federate. To avoid this problem, the federate must be modified so that RTI events that are directly transmitted between entities within the same federate be queued so that they are assigned the same time stamp as that selected within the interval. This could be accomplished by simply using the RTI to transmit events between entities within the same federate, or buffering these events within the federate until the final time stamp has been assigned. Note that this is only necessary for events passed between federates, i.e. it is not necessary for events that remain local to the federate.

Receiver-Independent Assignment Using ATC Synchronization

The algorithm that has been developed to realize the approximate time causal (ATC) mechanism provides a simple solution to the multiple destination message problem. This solution also applies to the synchronous version of the AT algorithm. ATC provides stronger ordering semantics than AT because it guarantees that causally related events are delivered in the correct order, even if their time intervals overlap.

In AT and ATC order, each event X is assigned a time interval [E(X), L(X)]. E(X) is called the E-time of the event, and denotes the earliest time event X could occur. L(X) is called the L-time, and denotes the latest time X could occur. In the ATC algorithm, each processor repeatedly cycles through the following steps:

1. Determine the minimum L-time called the ELT (earliest L-time) value among the unprocessed events in the simulation (e.g., see Figure 7).
2. Among those events whose time interval includes the ELT value, determine the largest E-time or LET (latest E-time) value (also shown in Figure 7).
3. Deliver events whose time interval overlaps with [LET, ELT]; each delivered event X may be assigned a precise time stamp within the interval [LET, L(X)].

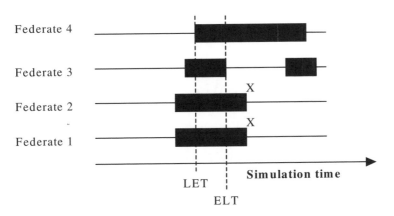

Figure 7. Snapshot of ATC computation.

The ATC algorithm guarantees that at the start of each cycle, all processors will have received messages for events that are distributed to multiple destinations.

Ensuring that multiple destination messages all receive identical time stamps (e.g., event X in Figure 7) is easy with this algorithm. One need only define a function f(X,LET,L(X)) that computes the assigned time stamp based on the contents of the event X, the computed LET value, and the L-time of the event. This function can be computed locally at each receiver. Because X, LET, and L(X) is the same at each receiver, each federate will compute the same time stamp on each processor.[1] The sender can also compute f(X,LET,L(X)), enabling it to return the assigned time stamp to the federate that scheduled the event without additional communication with other processors.

The ATC algorithm was implemented under a project sponsored by the DARPA Advanced Simulation Technology Thrust (ASTT) program. The results of the implementation, documented in [10], showed that ATC offers several advantages. It is simple, requiring no additional interprocessor communication to ensure consistent time stamps are assigned. It can be easily added to the ATC algorithm with minor modification. The only additional overhead is the computation time associated with executing the function f(X,LET,L(X)). However, it is anticipated that this will not require very much computation in most cases of practical interest.

Pre-Sampling Time Values

An alternate approach to solving the multi-destination problem is to use pre-sampling. The key idea in this approach is to pre-sample random numbers used to select time stamps within intervals, and to use this information to enhance the lookahead of the federate. The idea is as follows. Suppose the federate has zero lookahead without pre-sampling. If we know in advance that the time intervals used by the federate have size W and time stamps within this interval follow some probability distribution, we can sample the random number generator in advance, well before the event is generated. If the random number generator yields a value V, then this implies the next message generated by the federate must have a time stamp V units of simulation time in the future, i.e., the federate has a lookahead of V.

Unfortunately, there are several complications with this approach. Obviously, one must know the time interval size and probability distribution in advance in order to pre-sample the generator. More seriously, suppose the federate decided to send two messages rather than one after its next

---

1. modulo differences in floating point arithmetic in the different processors.

time advance. Then the lookahead would be the minimum of the two random number generator samples. What if the federate generated three? There needs to be some bound on the number of messages the federate will generate over its next time increment in order to correctly set the lookahead. Further, if this bound is high, the increased lookahead one can gain from this approach will be limited because the resulting lookahead will in general decrease as more random number generator samples are required.

A related question concerns future incoming messages and other local events. Suppose the federate advanced some small amount of time, then generated another set of new messages? These must also be pre-sampled, and a lookahead computed to take into account those potential messages.

In the following, we assume all events sent between federates utilize the same time interval and probability distribution to select the precise time stamp within the interval. To address the issues outlined above, the following federation specified parameters are defined:

- DT is a simulation time period; this can be set arbitrarily by the federate.
- M indicates the maximum number of messages the federate can produce over the next DT units of simulation time advance.

The federation must specify M in advance of the execution (although M could change during the execution) and adhere to this constraint throughout the execution. Violation of this constraint may result in a lookahead violation (generating a message with too small a time stamp).

Assume the federate has pre-sampled M random numbers at any instant in the execution. Let these samples be denoted $S[1], S[2], ..., S[M]$. The (enhanced) lookahead of federate i using time intervals and pre-sampling is:

$$L_i = L_i' + \min \{S[1], S[2], ... S[M], DT\} ,$$

where L_i' is the lookahead of federate i *without using time intervals.* Suppose the federate is currently at simulation time T. In the worst case, the federate will generate all M messages without advancing simulation time, so the minimum time stamp of any message it will produce among these M messages is $T + L_i' + \min \{S[1], S[2], ... S[M]\}$. However, once this has occurred, the federate is constrained to advance DT time units before it can generate any additional messages. Once it has advanced DT units, it is possible the federate could generate a message with time stamp $T + L_i' + DT$ since we have not pre-sampled the $M+1^{st}$ message. This is the reason DT is included in the lookahead equation.

Operationally, the pre-sampled random numbers can be stored in a circular queue. When a sampled value is needed, it is selected from the front of the queue, and a new random number is sampled and placed at the end, so the queue always contains M pre-sampled values. The enhanced lookahead must be recomputed at this point.

Experiments

An implementation of the pre-sampling algorithm described above was developed[2] in order to compare its performance with the ATC-order mechanism [10]. An RTI, called BRTI that implements variable lookahead and the ATC-RTI that implements ATC order were developed using version 2.3 of the RTI-Kit software package [12]. RTI-Kit is a collection of libraries implementing key mechanisms that are required to realize distributed simulation RTIs, especially RTIs based on the High Level Architecture. The RTI-Kit libraries are designed so they can be used separately, or together to enhance existing RTIs, or to develop new ones.

Applications

The experiments were conducted using well-known benchmarks widely used in the parallel discrete event simulation community. The first benchmark is the PHOLD synthetic workload program [13]. A single logical process is mapped to each federate, and LPs do not send messages to themselves. Thus, there are no local events, i.e., each event processed by a federate was generated on a different processor. The target time stamp is selected from a uniform distribution with mean 1.0, and the destination federate is selected from a uniform distribution. The minimum time stamp increment is zero, resulting in a zero lookahead simulation.

The second application is a queueing network (QNET), configured in a toroid topology. This simulation does *not* attempt to exploit lookahead. A textbook approach to realizing this simulation was used that includes both job arrival and job departure events. Each departure event schedules an arrival event at another queue with time stamp equal to the current simulation time of the LP scheduling the event, i.e., messages sent between LPs all contain zero lookahead. Service times are selected from an exponential distribution, and jobs are serviced in first-come-first-serve order. Each federate models a rectangular portion of the queueing network.

Execution Speed

The experiments described in [10] were repeated for both ATC-order and pre-sampling for both the PHOLD and QNET applications. As mentioned previously, all messages sent between federates use the same interval size and all simulations have zero lookahead. The pre-sampling experi-

---

2. This work was published in the 14th Workshop on Parallel and Distributed Simulation, Bologna, Italy, May 2000 [11].

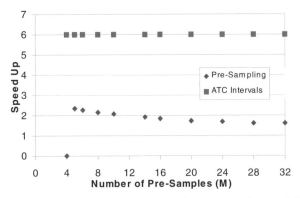

Figure 8. Speed Up of PHOLD simulation for the Pre-Sampling and ATC algorithms.

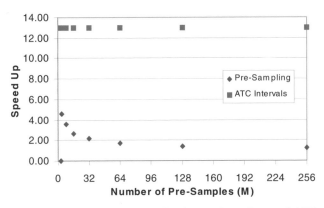

Figure 9. Speed Up of QNET simulation for the Pre-Sampling and ATC algorithms.

ments varied three parameters: INT, M, and DT. The data shown in Figures 8 and 9 are for DT = 0.1, INT = 0.5, and a variable M.

In Figures 8 and 9, the performance of the simulation is plotted relative to the distributed simulation using precise time stamps (and ATC-order) for the PHOLD and QNET benchmarks, respectively. As can be seen, pre-sampling provides speed up over the distributed simulation that uses precise timestamps. The speed up varies with the value of M. As M decreases, we see better speed up since there are fewer random number samples to select from. Therefore, there is a better chance of having a large S_{min}, which gives a larger lookahead and increases performance. However, M must be greater than 4 for this benchmark.

Also in Figures 8 and 9, we see the speed up of the pre-sampling approach compared to the ATC-order approach. The ATC experiments were run using the same interval size and federation message population. The ATC approach outperforms pre-sampling. This is because the increased parallelism using enhanced lookahead obtained by pre-sampling is not as large as that exploited by ATC.

Pre-Sampling Conclusions

In this research, we described a method of exploiting time stamp intervals using pre-sampling of random numbers to enhance lookahead. An advantage to exploiting temporal uncertainty in this manner is that it can be used on a conventional RTI using the existing RTI interface; no specialized time management algorithms are needed. Instead, two new functions are called by the federate to generate the pre-sampled values and replace the time stamp assigned to the outgoing message with one that includes the pre-sampled value. Another important advantage is that pre-sampling allows probability distributions to be used to define assigned time stamps. This offers the modeler greater flexibility and statistical control over the assigned time stamps.

The pre-sampling approach is relatively simple to implement, but requires several assumptions in order to realize the solution. These assumptions include fixed sized time intervals, the same distribution for all messages, and knowledge concerning the number of messages the simulation will produce in the "near-future" (M) before advancing simulation time. Pre-sampling does give performance improvement over a simulation that uses precise timestamps; however, lookahead enhancement is small if many random number samples are required.

Summary

This section described solutions to issues concerning messages with multiple destinations, and notifying the sender of a message of the time stamp assigned to an event. The receiver-independent assignment approach appears to be the most promising in that it provides a simple, straightforward solution to the multi-destination problem. Further, use of ATC order offers several advantages relative to AT-order, e.g., causally correct ordering of events and repeatable executions. Moreover, experimental data collected thus far indicates that ATC synchronization algorithm achieves nearly the same performance as the current AT synchronization algorithm implemented in the prototype. Thus the cost in terms of runtime overhead to provide the additional functionality of ATC order appears to be small, though additional experimentation and analysis are required. Thus, ATC order using receiver-independent time stamp assignment appears to be the preferred approach at this time.

CONCLUSIONS

Under this fellowship, I studied two outstanding issues related to AT and ATC order: Model Validation and Timestamp Assignment. I developed a methodology for Model Validation based on a simulation model's VV&A process. The steps that address time intervals were described as well as the specific V&V techniques to be used in the methodology. I implemented part of the process using a military federation called GTank. The results from the experiment showed that the sensitivity analysis is a good approach for examining the effect of the time interval on simulation results. For Timestamp Assignment, I described three solutions messages with multiple destinations. I implemented the pre-sampling approach and compare those results with an implementation of the receiver-independent assignment approach. The results showed that the receiver-independent assignment approach appears to be the most promising in that it provides a simple, straightforward solution to the multi-destination problem.

ACKNOWLEDGMENTS

I would like to gratefully acknowledge the Link Foundation for their support over the past year. I have made significant progress towards my proposal, which is planned for Spring 2001. This research resulted in refereed publication that will become part of my Thesis. Thank you for the opportunity.

REFERENCES

[1] Defense Modeling and Simulation Office (DMSO), "High Level Architecture Interface Specification," version 1.3 (1998).

[2] R. M. Fujimoto. "Exploiting Temporal Uncertainty in Parallel and Distributed Simulations." *Proceedings of the 13th Workshop on Parallel and Distributed Simulation*, 46–53 (1999).

[3] Defense Modeling and Simulation Office (DMSO), "Department of Defense Verification, Validation, and Accreditation (VV&A)" *Recommended Practices Guide* (1996).

[4] M.L. Loper, and R.M. Fujimoto, "Experience with Approximate Time Order." College of Computing, Georgia Institute of Technology: Atlanta, GA (2000).

[5] R.E. Shannon, *Systems Simulation: The Art and Science*. Englewood Cliffs, NJ: Prentice-Hall (1975).

[6] A.M. Law and W.D. Kelton, *Simulation Modeling and Analysis (2nd ed.)*. New York, NY: McGraw-Hill (1991).

[7]   C.F. Hermann, "Validation Problems in Games and Simulations with Special Reference to Models of International Politics." *Behavioral Science, 12*(3): 216–231 (1967).

[8]   D.R. Miller, "Sensitivity Analysis and Validation of Simulation Models." *Journal of Theoretical Biology, 48*(2): 345–360 (1974).

[9]   R.L. Van Horn, "Validation of Simulation Results." *Management Science, 17*(5): 247–258 (1971).

[10] R.M. Fujimoto, *Approximate Time Causal Order.* College of Computing, Georgia Institute of Technology: Atlanta, GA (1999).

[11] M. Loper, and R. Fujimoto. *Pre-Sampling as an Approach for Exploiting Temporal Uncertainty. Proceedings of the 14th Workshop on Parallel and Distributed Simulation.* Bologna, Italy: IEEE Computer Society, p. 157–164 (2000).

[12] R.M. Fujimoto, and P. Hoare. *HLA RTI Performance in High Speed LAN Environments. Proceedings of the Fall Simulation Interoperability Workshop.* Orlando, FL (1998).

[13] R.M. Fujimoto, *Performance of Time Warp Under Synthetic Workloads. Proceedings of the SCS Multiconference on Distributed Simulation* (1990).

Research Status of Context Recognition for Synchronization of a Behavioral Vehicle Model for Embedded simulation

William J. Gerber

School of Electrical Engineering and Computer Science

University of Central Florida

P.O. Box 160000

Orlando, FL 32816

Research Advisor: Dr. Avelino J. Gonzalez

ABSTRACT

Embedded simulations hold great promise for highly realistic training at reduced costs by combining live, virtual and constructive simulations, but they have technical challenges to overcome. One of those, the limitation on available bandwidth to meet the huge communication demands of embedded simulation, is being addressed through research on the use of behavioral models. These models are intended to improve on current dead reckoning modeling techniques to reduce the communication bandwidth needed. This is accomplished by synchronizing the behavioral models of human-controlled vehicles with the actual vehicle such that the models can be accurate for longer periods of time. The synchronization is done through the use of context-based representation. The behavioral model performs the actions that are appropriate for the behavioral context of the actual vehicle it represents while the model is on the other vehicles in an embedded simulation. Those actions for a particular context would have been learned previously by observation of human actions and the neighboring environment when the human, while in that behavioral context, was taking those actions. The behavioral model incorporates inputs of the surrounding environment so that the model dynamically responds as the observed human would respond in the same situation. However, the model has to know what the current behavioral context is in order to respond with the correct actions. This paper will report on the status of the research at the University of Central Florida into using template-based reasoning, neural networks and learning-by-observation to recognize the current behavioral context so that it can be used for synchronizing vehicular behavioral models in embedded simulations.

INTRODUCTION

Background

Simulations, whether constructive, virtual or live, have become valuable tools for individual and group training. They generally allow acceptably realistic training at a lower cost than using the actual equipment in a fully operational capacity. Additionally, the use of distributed simulations has allowed for team and large scale unit training in a constructive or virtual environment without the expense of massive movements of equipment to a central training area and the scheduling restrictions of that limited training area resource. Furthermore, the introduction of Computer Generated Forces (CGF's) has added greater realism into the simulation by allowing representations of the enemy forces as well as the inclusion of greater numbers of simulated friendly forces than can be reasonably assembled for any given training exercise.

The most realistic training, however, occurs when the individuals and crews can train on their own operational equipment in the actual environment. This is done now in live simulations using instrumented large-scale exercises at locations such as the U. S. Army's National Training Center. There, the live fire of weapons is replaced for increased safety and reduced expense by the use of instrumented vehicles where the outcome of individual engagements is determined by computer processing of the data in real time. The initial data and the results are transported over a communications network, the Range Data Measurement System (RDMS), which allows only 2400 bits per second peak bandwidth per vehicle for the information exchanges between vehicles. [1] The time lags and bandwidth of the communications network, though, still restrict the realism.

Embedded Simulation

A combination of live, virtual and constructive distributed simulations that is highly promising for greater realism in training at reduced costs, called embedded simulation, is being explored for use in combat vehicles by a program of the U. S. Army's Simulation, Training and Instrumentation Command (STRICOM). That program is called the Inter-Vehicle Embedded Simulation Technology (INVEST) Science and Technology Objective (STO) program. [2] Among the many technical challenges the program must overcome is that of providing a simulation environment in which live vehicles, manned vehicle simulators, and computer generated forces can interact with each other as well as with the battlefield environment in real-time over a geographically diverse, distributed network.

Behavioral Vehicle Model

A significant problem is the high communications requirements imposed by the need to convey large amounts of data among the various players. The Vehicle Model Generation and Optimization for Embedded Simulation (VMGOES) project at the University of Central Florida is focusing on this aspect of the INVEST STO program. [3,4,5] The approach is to use a behavioral vehicle model (VM) that is context-based to match the actions of each human-controlled entity on the battlefield. [5,6] That model would be resident on each vehicle that could interact with it. By observing the surrounding environment of each VM's location in the simulation at each update time step, each VM will perform the actions that are appropriate for its current behavioral context. This will allow each VM to match its human-controlled entity's behavior for a longer period of time than is possible with only dead-reckoning updates, thus reducing the communications bandwidth required.

However, discrepancies between the VM and the human controlled entity it models will inevitably occur and these must be detected and resolved to allow the VM to function efficiently. The portion of the model that addresses this need, the Difference Analysis Engine (DAE), will be resident only on the human-controlled entity. Figure 1 provides a functional dia-

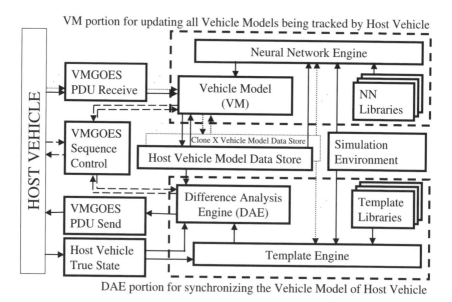

Figure 1. VMGOES Architecture with the interactions for the Vehicle Model and Difference Analysis Engine portions on the Host Vehicle shown.

gram of the VMGOES architecture, illustrating the VM and DAE portions. The DAE will be able to observe the actual host vehicle's actions as well as the simulation environment and uses its host vehicle VM as a reference. It then must evaluate, at each update time step, whether a significant discrepancy exists between its VM and its actual vehicle's state. If a significant discrepancy does exist, it will immediately take the action needed to synchronize each VM representing its host vehicle with its actual host entity's state through communicating information packets called Protocol Data Units (PDU's) to each of the other vehicles using its VM. This corrective action can involve (1) a simple State Realignment PDU to update the vehicle model's location, direction and speed; (2) a forced VM Context Shift PDU to match the context of the human-controlled entity; (3) a Model Correction PDU to change the way the model itself responds; or, as a last resort, (4) a Model Suspension PDU to revert to standard dead-reckoning until the DAE can recognize what context the human-controlled vehicle actually is in. The information included in the State Realignment PDU is also included in the other three correction types and none of the corrections are made unless the VM exceeds a set error bound for location and/or orientation which would require a PDU to be sent. A more detailed description of the architecture and operation of the VMGOES behavioral model was presented in a paper at the Interservice/Industry Training, Simulation and Education Conference 1999. [7] This paper, however, will focus on the status of the research into the DAE function of Model Correction and on how techniques, such as temporal template based reasoning and neural networks, are being used to accomplish that DAE function.

RESEARCH

In describing this research, background on template-based reasoning and how learning-by-observation will be used with neural networks to create those templates will first be covered. Then, the means of using ModSAF to substitute for the human controlled vehicle in testing the concept for behavioral context recognition will be discussed. Finally, the process of using learning-by-observation for developing the template attributes for context recognition and the experimental results to date will be presented.

Template-Based Reasoning

At the center of the model synchronization research is the challenge of determining the actual entity's intentions/behavioral context. Temporal template-based reasoning is being explored for this function, where each defined intention/behavioral context that the vehicle could assume is as-

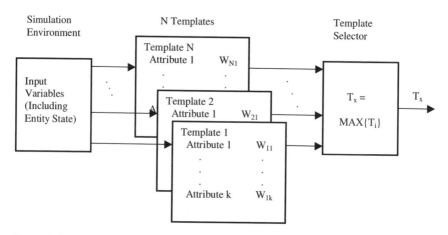

Figure 2. Temporal Template Components. The score for each template is either the sum of the weighted values for the attributes or else zero if the sum is not above a critical threshold value. The template selected as the winner is the one that has the largest score.

sociated with a template. [8,9] Each template would have selected attributes that represent portions of the vehicle's state and simulation environment (see Figure 2). The set of attributes selected need include only those germane to the determination of that particular template's validity. At each update cycle, each template's attributes would be evaluated and multiplied by a weighting value to determine an overall score for the template. If a minimum critical threshold value were not reached, the template would not be considered further during that update cycle for competition as a candidate for the vehicle's intention/behavioral context. The template chosen from among the competing templates as the one representing the actual vehicle's intention/behavioral context would be the one with the highest score.

Automating Template Training by Observation

Closely associated with the use of the templates is the challenge of how to automate the setting of the attribute weights. In earlier research with temporal templates, the values for the weights were manually determined by having the researcher adjust the values until the results were satisfactory. [9] That was possible for the limited domain that was involved. However, for a much larger domain, such as for the ground vehicle in a battlespace with many possible courses of action, a practical method of automating the setting of the weights by observation of representative examples was essential. This not only would allow the determination of the

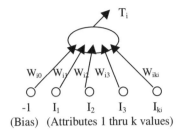

T_i

W_{i0} W_{i1} W_{i2} W_{i3} W_{iki}

-1 I_1 I_2 I_3 I_{ki}
(Bias) (Attributes 1 thru k values)

Figure 3. Attribute Learning by Observation Using Neural Network Framework.

weights, but would also allow the templates to be readily created to match various vehicle entities whose actions might differ in the same circumstances. For example, generalized templates could be tailored for individuals/crews whose training levels are expert, average or novice and specialized templates could be created to represent a specific individual/crew for the most efficient representation.

Although a large number of observations could be made and data collected, the question would still remain about how to use that data for setting the attribute weights. Observing the structure of the templates, one could make the analogy that each template is similar to a single layer neural network with a linear activation function, i. e., the set of inputs (attributes) are each individually multiplied by their own weights and the result is summed to produce an output (see Figure 3). That being so, a neural network training procedure, such as back-propagation, could be used to set the value of the weights. In this case, the template score, the sum of each input attribute multiplied by its associated weight, corresponds to the output of the neural network with a linear activation function. It was therefore postulated that neural networks could be used for determining these weights and, in addition, the templates would not have to be restricted to a single layer. Multiple layers would allow for discrimination between more complex behaviors, since a single layer neural network can only be trained to discriminate between linearly separable groupings.

ModSAF Entity Data Substituted for Human Controlled Entity Data

At the start of this research, a simulator was not available to allow collection of data on a human-controlled entity. Additionally, the VM and DAE were being developed within a ModSAF environment. Therefore, the decision was made to use a ModSAF entity as a means to collect data that would be similar to the type of data that would be collected from a human-controlled entity. Since the INVEST STO was addressing the M1A2 SEP vehicle, the M1A2 ModSAF entity was chosen to substitute for the human-controlled entity. Furthermore, to scope the investigation, the entity was given the order to execute a Road March on a single selected section of road on the National Training Center database (NTC-0101). That route covered approximately 7 kilometers and included 45 separately defined road segments with 17 turns of more than 3 degrees. Each road segment is a straight line defined by a waypoint at each end where it joins other road segments. For this initial research, a single turn was selected for detailed analysis. It was a left turn of 56.8 located approximately 3 kilometers from the start of the route.

Behavioral Sub-contexts Determined for Road March Context

To see if behavioral sub-contexts could be observed for how the ModSAF M1A2 entity maneuvered at the selected turn, 110 runs were made on the route using ModSAF in a non-repeatable mode. The initial starting location was the same for each run at the beginning of the route, but eleven initial headings were used with 10 runs starting at each of the different headings. Those headings varied in 1 increments from 6 left to 4 right of the initial road segment heading.

A trace of the path at the selected turn was printed for each of the 110 runs and visually categorized into separate behaviors for the approach and departure portions of the turn. All the paths started approximately on the centerline of the road segment for the approach and went to the right of the waypoint defining the intersection of the two road segments. The initial moves away from the centerline on the approach were categorized as Early (62 were more than 45 meters before the waypoint), Nominal (41 were between 30 and 45 meters) or Late (7 at less than 30 meters). Fifteen of the Early approach types exhibited a double curve behavior where the entity initially swung out to the right, then turned back slightly to an intercept heading, and finally continued back away from the centerline to pass to the right of the waypoint. These behaviors could be described easily with simple rules and did not need anything more complicated to distinguish between those types of approaches to the turn. They could be represented as four competing templates: (1) Early, Single Curve; (2) Early, Double Curve; (3) Nominal, Single Curve; and (4) Late, Single Curve.

However, the departures after passing the waypoint were more complicated and could not be described by simple rules. The categorization of the departure patterns resulted in the 10 types shown in Table 1 with the number of runs and description of each type listed.

Only the first six types, with seven or more examples for each, were considered for training of neural network weights for departure type competing templates. Since the last four types had no more than three examples of each, they were grouped together as a category called Other (OT) and no

Table 1. Departure Pattern Types

Departure Types	Number of Runs	Description
Flat Nominal (FN)	15	Parallels outbound segment, then turns to intercept at about 45 meters beyond the turn (See note below)
Flat Bow Low (FBL)	16	Same as FN, but slight bow before turn to intercept
Flat Bow High (FBH)	17	Same as FBL, but slightly more pronounced bow
Bow Nominal (BN)	35	Definite bow on the departure that continues to an intercept at about 45 meters beyond the turn
Straight Angled (SA)	10	After initial turn, maintains an intercept heading with little or no change to an intercept beyond 80 meters
Bow Wide (BW)	7	Much wider bow than BN with intercept beyond 80 meters
Bow Asymptotic (BA)	3	Initially similar to BN, then smoothly reduces intercept heading to asymptotically intercept beyond 60 meters
Bow Distant (BD)	3	After initial turn, slowly turns to an intercept heading beyond 80 meters resulting in an extended slight bow
Double Curve (DC)	3	Initially turns to a heading that would intercept close to the turn, then turns away from the segment followed by a turn back that closely resembles the SA pattern
Close Intercept (CI)	1	Intercepts within about 10 meters followed by a long-lasting overshoot before eventual reintercept

Note: Nine or the FN departures initially turn to a sharp intercept heading close to the turn, then turn to the parallel heading.

attempt was made to train templates to recognize them. To account for the times when the templates could not recognize the departure type or for the cases when none of the templates were applicable, two more templates were developed for the Road March context to differentiate between the straight portions of the route and the turns. They were simple rule-based templates that considered entity locations within 25 meters of a turn as Generic Turn and all other locations as Generic Straight. The output value of each of these two templates was set to be less than the threshold values of the templates for the approach and departure types. Thus, whenever the major context was Road March, at least one of the Road March Sub-context competing templates would be a winning template.

Template Attributes Selected for Road March Sub-contexts

The attributes that were to be used included both Mandatory and Independent attributes. [5] The Mandatory attributes included Road Following, where the entity had to be within 5 meters of the roadway centerline, and other attributes that had to be triggered (in the correct sequence in some cases) before the various approach and departure types could be considered. These were included in Support Templates that covered such conditions as approaching the turn, being near the centerline, turning away from the centerline to the outside of the curve, etc. The Mandatory attributes served to only allow the template to be considered for competition but did not contribute to the template's competitive value.

The Independent attributes were those qualities of the entity's state and its relationship to its environment that could be observed and were thought to possibly contribute to the recognition of the entity's intentions. Five different attributes were selected and are defined in Table 2, where Waypoint A is the last waypoint passed and Waypoint B is the current waypoint that the entity is attempting to reach. Since road segments are defined by the locations of the waypoints at each end, when ModSAF considers that the entity is past a turn, the coordinates for Waypoint A and Waypoint B change to reflect the new current segment.

These attributes were considered in setting up the templates, both Competing and Support. In fact, the two categories of neural networks that were trained for recognition of the entity's departure type intentions were treated as Support Templates. They were named "History" and "Recent" for the inputs that were used and which will be described shortly. Each of the six departure types trained had its own set of History and Recent Support Templates. The outputs from the History and Recent Support Templates, which were basically the outputs of the trained neural networks clipped to remain between plus and minus one, were treated as input attributes to

Table 2. Independent Attributes

Attribute Name	Description
DA	Distance from Waypoint A, as projected onto the centerline from Waypoint A to Waypoint B.
DB	Distance from Waypoint B, as projected onto the centerline from Waypoint A to Waypoint B.
XD	Cross distance, as measured perpendicularly from the centerline between Waypoints A and B.
HR	Heading of the entity relative to the direction of the centerline from Waypoint A to Waypoint B.
SP	Scalar speed of the entity across the surface.

their respective departure Competing Template. Finally, in the Competing Templates, the attribute values from the History and Recent Support Templates were to be mixed using Certainty Factors to provide the output value for the Competing Templates.

To keep the neural networks tractable in size and the data manipulation and storage requirements at a reasonable level, inputs for XD, HR and SP were sampled and stored 26 times for each departure portion of the turn segment. Specifically, those three inputs were sampled at the first update cycle after the entity changed from approaching the current turn's waypoint to attempting to reach the next waypoint. Similarly, the three inputs were sampled each time the entity's value for DA went past a two meter increment, i. e., when it first passed two meters, four meters, six meters, etc. up to having passed 50 meters.

For the History Support Templates, the 78 values for XD, HR and SP at all 26 possible sample times were used with zeros substituted for the values when they had not yet been sampled. Thus, "history" was used as the name for these neural networks/templates because every value that had been sampled to that point since the entity passed the current turn's waypoint was used as an input. For the Recent Support Templates, however, only a certain number of the most recently sampled values were used as inputs along with a DA input that indicated how far past the waypoint the most recent values were sampled. While various numbers of recent values were tried parametrically, along with whether or not to include the DA input, the use of the last four samples of XD, HR and SP and the most recent DA input was decided upon as the values to use for inputs during training and operation. Thus, the "recent" neural networks/templates had a total of 13 values for XD, HR, SP and DA. Zeros were substituted for the oldest values of XD, HR and SP that didn't exist until the entity had passed the point where DA exceeded six.

Fuzzy Gaussian Transformation of Raw Data for Neural Network Inputs

To normalize the input values for back-propagation training of the neural networks, a fuzzy transformation was effected upon each of the input values for the particular departure type being trained The normalized inputs thus indicated the fuzzy membership of the data to the particular departure type. The transformation was based on the Gaussian Probability Density Function. Determining the mean and standard deviation values for each of the XD, HR and SP inputs at each DA value for the examples that were of that departure type was accomplished. That resulted in a transformation data set of 26 values for the mean and 26 values for the standard deviation for each of the three input values for each of the six departure types.

Where m_{DA} is the sample mean and s_{DA} is the sample standard deviation for the departure type input value at a particular DA value, the fuzzy Gaussian transformation equation used to produce the normalized fuzzy value $ff(y_{DA})$ from the raw input value y_{DA} was:

$$ff(y_{DA}) = e^{-(z^*z/2)}, \text{ where } z = (y_{DA} - m_{DA})/s_{DA}.$$

The raw data was extracted from each of the 110 runs so that there were 26 examples for each run for the History neural networks with each succeeding example having three more data items filled in over the initial zeros. The 26th example would have all 78 data items filled in over the zeros. To provide the training, validation and testing data sets for each of the departure types, the raw data for each run was normalized. The fuzzy Gaussian transformation equation and the transformation data set for each departure type was applied so that each run had six normalized, transformed data sets, one for each departure type. At the same time, the desired output value for each run was added at the end of each example—a one if the run was of the departure type whose transformed data set was used or a zero otherwise.

For the Recent neural networks, the normalized, transformed data sets were extracted from the History data sets as a subset with some minor manipulations required. For the first three examples for each run, zeros had to be added to represent the inputs for the second, third and fourth most recent sets of values that did not yet exist. In performing the manipulations for all the examples, all the inputs past the DA value being considered were deleted, the DA value divided by 100 was added as a last input just prior to the output value and, additionally, all outputs that were 0 were changed to –1.

Division of Input Examples into Training, Validation and Testing

In dividing up the examples from the 110 runs between training, validation and testing, an arbitrary number of 15 percent each was selected for validation and testing with the remaining 70 percent left for training. Table 3 shows the specific division of the number of runs for each based on the departure types available.

To decide which specific runs within each of the six departure types to select for training, validation and testing, the average value of the 26th example for each run was evaluated. That example was selected because all the data items would be filled in. The normalized, transformed data set used for each run was the one that resulted from using its own transformation data set. Therefore, the average of the 26th example data items was a good measure of the fuzzy membership of that run within its own depar-

Table 3. Numbers of Runs for Training, Validation and Testing by Departure Type

Departure Type	Raw Number	Trn(70%)	Val(15%)	Tst(15%)	Trn(70%)	Val(15%)	Tst(15%)
		Fractional Numbers			Final Selected Numbers		
BN	(35)	24.5	5.25	5.25	25	5	5
FBH	(17)	11.9	2.55	2.55	12	2	3
FBL	(16)	11.2	2.4	2.4	11	2	3
FN	(15)	10.5	2.25	2.25	11	2	2
SA	(10)	7.0	1.5	1.5	6	2	2
BW	(7)	4.9	1.05	1.05	5	1	1
OT	(10)	7.0	1.5	1.5	6	2	2
Total	(110)	77.0	16.5	16.5	76	16	18

Table 4. Membership Rankings of Runs for Validation and Testing by Departure Type

Departure Type	Raw Number of Runs	Ranks of Validation Runs	Ranks of Testing Runs
BN	(35)	6, 12, 17, 23, 29	5, 11, 18, 24, 30
FBH	(17)	7, 11	5, 9, 13
FBL	(16)	6, 10	4, 8, 12
FN	(15)	7, 9	6, 10
SA	(10)	4, 7	3, 8
BW	(7)	5	3

ture type. All the runs were then rank ordered within their own departure type and the runs for validation and testing were selected. To avoid extreme examples and to spread out the runs across the rankings of membership examples, the following rough selection criteria were used. For selecting one run, take one near the middle ranking. For two runs, take one a third of the way away from each end. For three or four runs, take one a fourth of the way away from each end and the remaining one or two from the middle or closely bracketing the middle. For five runs, take one each a sixth and a third from each end and the fifth from near the middle. The rankings for the runs selected for validation and test are shown in Table 4. Although the specific runs were selected based on their fuzzy membership rankings within their own departure type, the same specific allocation of runs were used for training, validation and testing for all six departure type Support Templates/neural networks.

Training of "History" Neural Networks

The back-propagation neural network training program used was developed by the researcher for his master's degree thesis, where it is de-

Table 5. Statistical Summary of Mean Absolute Errors for 60 History
Neural Networks

Depart. Type	Training Epochs		Training MAE		Validation MAE		Testing MAE	
	Average	StdDev	Average	StdDev	Average	StdDev	Average	StdDev
BN	144,565	35,140	.0444	.0005	.0962	.0005	.1345	.0009
FBH	95,545	6,493	.0153	.0004	.0048	.0004	.0733	.0002
FBL	97,605	66,071	.0291	.0016	.0176	.0006	.0199	.0012
FN	22,690	2,188	.0279	.0008	.0814	.0017	.0846	.0018
SA	183,175	57,553	.0112	.0017	.0648	.0032	.0150	.0011
BW	141,335	34,044	.0229	.0021	.0292	.0023	.0182	.0019

scribed in detail with the source code in an appendix. [10] The program allows the researcher to set the training rate, momentum factor, maximum number of epochs to allow, seed for random initialization of the weights, the threshold value for the average training error that must not be exceeded before an increase in the validation error will be allowed to stop the training, and the number of epochs to skip between checks for validation error increases. The activation function is linear for the output node(s) and sigmoidal for the hidden layer nodes.

Initially, a brief parametric analysis was undertaken to judge what affect different values of training rate, momentum factor and number of hidden nodes might have on the quality of the neural network produced. The data set for FBL was used. The number of hidden nodes were varied (1, 5, 10, 20, 39 and 78) with the training rate set at 0.01 and momentum factor set at 0.9. One seed was used for training one neural network for each setup. Similarly, eight additional networks with 10 hidden nodes each were trained so that all the combinations of three training rates (0.001, 0.01, 0.1) and three momentum factors (0.0, 0.5, 0.9) were used. The validation threshold was set at 0.14 for the average training output error before an increase in the average validation output error would stop the training. It appeared that a training rate of 0.001 with 0.0 for the momentum factor was the better set of parameters to use, particularly for discrimination at higher cutoff values. The data indicated that the number of hidden nodes should be low (1 or 5).

Next, two sets of neural networks were trained for each of the six departure types using a training rate of 0.001 and zero momentum. One set had five hidden nodes and the other set had one hidden node. As the results were very similar, nine more neural networks for each departure type were trained using the single-hidden-node architecture. The statistical summary of the training of the 60 single-hidden-node neural networks is shown in Table 5.

From Table 2, it can be seen that within departure types, there is no fixed order for whether training, validation or testing mean absolute errors

(MAE's) will be the highest or lowest. For example, the training MAE is the highest within one departure type (FBL) and the lowest within three (BN, FN, SA); the validation MAE is the highest within two (SA, BW) and the lowest within two (FBH, FBL); and the testing MAE is the highest within three (BN, FN, FBH) and the lowest within one (BW). It is also of note that the random selection of initial weights does not significantly affect the final error results for the capability of the neural networks trained with a given data set of examples, as can be seen by the relatively small standard deviations compared to the averages.

Results of "History" Neural Networks for Behavior Recognition

Since the same ten seeds were used for training the one-hidden-node neural networks for each departure type, each of the sets of six departure neural networks trained with the same seed was arbitrarily grouped together. Each of these ten groupings was then competed internally using the 18 test runs that had not been used in training or validation. The resulting 468 test points were evaluated for which neural network (Support Template equivalent) had the highest output and whether or not it met the Critical Threshold (Tc) requirement to be declared a winner. If the departure type template that won matched the test run departure type, it was counted as a "Correct ID". If the winning departure type template did not match the test run departure type, it was counted as a "False ID". Finally, if no departure type template won, i.e., none was above the Tc requirement, it was counted as a "No ID". The Correct ID's for the ten groupings are plotted in Figure 4 against Tc values from 0.0 to 0.5 while the False ID's and No ID's are similarly plotted in Figure 5. In Figure 5, the monotonically decreasing (as Tc decreases) set of ten lines is for the No ID's. Note that there is little variation between the results from the ten groupings of individually trained neural networks. This would indicate that the results are relatively insensitive to the randomness of the initial weights for training—any group of neural networks trained on the same data produce approximately the same result.

From Figure 4 it can be seen the number of Correct ID's continues to increase as the Tc decreases. It does this with an increasing rate below a Tc of 0.2 where the percentage of Correct ID's increases from the 76.7–78.2% range to the 82–82.7% range. At a Tc of 0.5, the range was 74.4–74.8%. From Figure 5 it can be seen that the percentage of False ID's only slightly increases from a Tc of 0.5 (12.0–12.4%) down through a Tc of 0.05, where the rate is 12.4–12.8%. However, the percentage of False ID's shoots up to 17.3–18.0% at a Tc of 0.0. This would indicate that using a Tc of less than 0.05 would greatly increase the False ID's while only marginally increasing the Correct ID's.

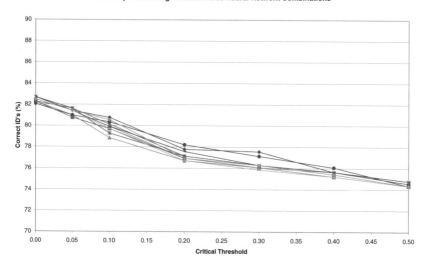

Corrrect ID's vs Critical Threshold
10 Independent Single-Hidden-Node Neural Network Combinations

Figure 4. Correct Recognition Results for "History" Neural Networks vs. Critical Thresholds.

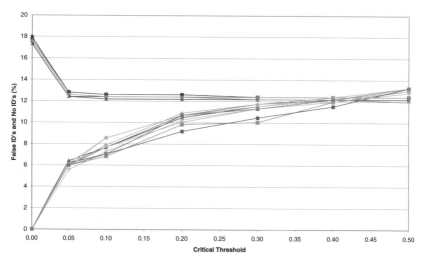

False ID's and No ID's vs Critical Threshold
10 Independent Single-Hidden-Node Neural Network Combinations

Figure 5. False or No Recognition Results for "History" Neural Networks vs. Critical Thresholds.

The six neural networks that were trained with five hidden nodes were evaluated similarly for comparison but only for a Tc of 0.1 and 0.5. At a Tc of 0.5, the five-hidden-node neural networks had False ID rates (12.4%) very similar to the one-hidden-node neural networks and had a better Correct ID rate (81.0% vs. 74.4–74.8%). However, at a Tc of 0.1, the five-hidden-node neural networks had a Correct ID rate of 82.3% and a False ID rate of 17.7%. This was only marginally better than the one-hidden-node neural network for Correct ID's (82.3% vs. 78.9–80.8%) but considerably worse than the rate for False ID's (17.7% vs. 12.2–12.6%).

Another observation made was that the neural networks were relatively insensitive to the values input in the latter stages of the departure. Thus, if a departure started out similar to a BN but later deviated from the pattern, such as with a BA, the History Support Template would most likely still provide a high value for BN. The idea of the Recent Support Template neural network was developed to try to offset that weakness of the History Support Template.

Training of "Recent" Neural Networks

As with the History neural networks, a limited parametric study was performed to first determine the architecture to use. For the first parts of the study, only the two most recent sets of inputs were used as well as a DA input, so the input layer had only seven input nodes. Two, one and no hidden layers were used. The three single-hidden-layer architectures had one, five or ten nodes while the four two-hidden-layer architectures had five or ten nodes in the first hidden layer and one or five hidden nodes for the second hidden layer. The output layer had only one node, of course, to produce the desired output (plus or minus one). The architecture selected as a result of the study had two hidden layers with ten and one nodes for the first and second hidden layers, respectively.

Next, a parametric study was undertaken to choose between one, two, three or four most recent sets of three inputs and with or without the DA input. The conclusion reached was that including the four most recent sets of three inputs and the DA input provided the best results.

Finally, a parametric study using the 13–10–1–1 architecture (13 input nodes, 10 first-hidden-layer nodes, one second-hidden-layer node, and one output node) was undertaken to determine what values of training rate and momentum factor to use. Initially, momentum factors of 0.0, 0.3, 0.6 and 0.9 were used as were training rates of 0.01, 0.0033 and 0.001. A relationship also was noticed that, for some of the departure types (BN, FBH), the maximum output values peaked below the desired 1.0 level and varied in the opposite direction of both the training rate and momentum factor. That is, as either or both increased in value, the maximum output value

Table 6. Training Parameters Selected for Recent Neural Network Training

Depart. Type	Training Rate	Momentum Factor	Validation Threshold
BN	0.0010	0.0	0.4
FBH	0.0010	0.3	0.1
FBL	0.0033	0.3	0.1
FN	0.0033	0.3	0.2
SA	0.0001	0.3	0.2
BW	0.0010	0.3	0.1

produced tended to decrease. This indicated that for some of the departure types, smaller values of training rate and momentum factor should be considered. The scope of the study was then expanded for all six departure types to cover training rates of 0.0033, 0.001, 0.00033 and 0.0001 and momentum factors of 0.0 and 0.3. Unlike the History neural networks, it was apparent that different departure types would require different training parameters for the Recent neural networks. Table 6 provides the values selected, as a result of the study, for training rate, momentum factor and validation threshold to train the Recent neural networks.

Using the parameters from Table 6 with a neural network architecture of 13–10–1–1, sixty neural networks were trained using the data sets for Recent examples. The statistical summary of the Recent neural network mean absolute errors (MAE's) for training, validation and testing are shown in Table 7.

Comparing Table 7 for the Recent neural networks with Table 5 for the History neural networks, it is obvious that the errors for the Recent neural networks are consistently larger. That is true even after adjusting for the fact that the range of output values for the Recent neural networks is twice that of the History output values. The only exception is for the FBH testing errors where the Recent MAE is slightly less.

Table 7. Statistical Summary of Mean Absolute Errors for 60 Recent Neural Networks

Depart. Type	Training Epochs Average	Training Epochs StdDev	Training MAE Average	Training MAE StdDev	Validation MAE Average	Validation MAE StdDev	Testing MAE Average	Testing MAE StdDev
BN	40,925	18,947	.2967	.0447	.3562	.0288	.3567	.0388
FBH	35,880	23,779	.0343	.0116	.0180	.0077	.0707	.0111
FBL	16,570	14,377	.0363	.0147	.0326	.0139	.0635	.0143
FN	41,285	24,588	.0898	.0268	.1388	.0153	.1520	.0178
SA	75,330	24,972	.1043	.0052	.0891	.0072	.0838	.0056
BW	7,010	1,652	.0427	.0041	.1122	.0046	.0406	.0032

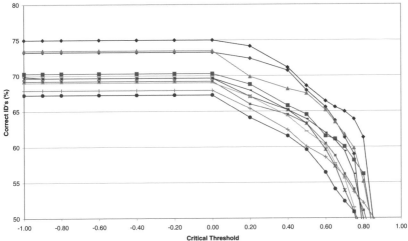

Figure 6. Correct Recognition Results for "Recent" Neural Networks vs. Critical Thresholds.

Results of "Recent" Neural Networks for Behavior Recognition

As with the History neural networks, the ten Recent neural networks for each departure type were apportioned to form ten groupings of six competing templates. Each grouping was competed internally and the results for the behavioral context recognition were plotted against Tc, where the range of Tc was from -1.0 to $+1.0$. The Correct ID's are plotted in Figure 6 and the False and No ID's are plotted in Figure 7. In Figure 7, the monotonically decreasing (as Tc decreases) set of ten lines is for the No ID's.

From Figure 6, it can be seen that the percentage of correct ID's increases to the range of 67.3–75.0% as Tc decreases to 0.0 and then remains essentially constant thereafter for each grouping as Tc decreases to a value of -1.0. (The only exception among the groupings is that one increases by 0.21% as Tc decreases from -0.9 to -1.0, the equivalent of one more correct ID.) The range of Correct ID's between the different groupings is a 7.7% spread, an order of magnitude more than the 0.8% spread for the History results at a Tc of 0.05. One can generalize that any Tc of 0.0 or less will give the maximum percentage of Correct ID's.

From Figure 7, the percentage of False ID's has a pattern similar to that for the Correct ID's. The False ID's percentage increases to the range of 9.2–11.3% as Tc decreases to 0.0 and remains constant after that as Tc continues to decrease to -0.9. However, as Tc decreases to -1.0, the range of False ID's

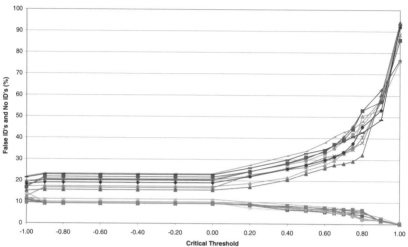

False ID's and No ID's vs Critical Threshold
10 Independent 13-10-1-1 "Recent" Neural Network Combinations

Figure 7. False or No Recognition Results for "History" Neural Networks vs. Critical Thresholds.

jumps to 9.6–14.3%. One can generalize that any Tc between 0.0 and –0.9 will not increase the number of False ID's.

The minimum rate of increase for Correct ID's for a Tc greater than 0.0 is in the range of Tc between 0.0 and 0.5, where the Correct ID's percentage is 59.6–68.6%. Thus the Correct ID percentage increases as Tc decreases from 0.5 to 0.0 by 6.4–7.6%. In the same range of Tc values, the False ID's percentage increases from 6.0–8.6% by 3.2% to the range of 9.2–11.8% as Tc decreases. Thus, decreasing Tc from 0.5 to 0.0 causes the percentage of Correct ID's to increase at more than twice the rate of False ID's.

Recall that one of the objectives for developing Recent neural networks was to offset the inertia of the History neural networks whereby an early identification is unlikely to be reversed if the entity's path departs from the identified departure type at the later stages of the departure. Therefore, the best Tc to use for the Recent Support Templates would seem to be at –0.9 so that strong negative values from the Recent Support Templates could offset lingering positive values from the History Support Templates.

SUMMARY

For the limited actions investigated to date, it appears possible to automate the setting of attribute weights for template-based behavioral context recognition, used in synchronizing behavioral models in embedded simulations, through the use of neural networks trained with observed examples of behavior. Research with neural networks, trained on categorized examples of the actions of an entity conducting a Road March on one turn, show that they can correctly identify the specific, categorized action being taken at more than an 80 percent rate while falsely identifying at a less than 13 percent rate. Research is still ongoing on possible synergistic effects of combining the results of the History and Recent Support Templates and on incorporating them into a behavioral model testbed in a ModSAF simulation.

ACKNOWLEDGEMENTS

This researcher was assisted in his efforts by a Link Foundation Fellowship and by an Interservice/Industry Training, Simulation and Education Conference Scholarship. The work itself was sponsored by the U.S. Army Simulation, Training, and Instrumentation Command as part of the Inter-Vehicle Embedded Simulation and Technology (INVEST) Science and Technology Objective (STO), contract N61339–98–K-0001. Those multiple means of support are gratefully acknowledged.

REFERENCES

[1] H. Bahr and R. F. DeMara, "A concurrent model approach to scaleable distributed interactive simulation", Proceedings of the 15th Annual Workshop on DIS, Institute for Simulation and Training, (1996).

[2] H. Bahr, "Embedded simulation for ground vehicles", *Proceedings of the Spring 1998 Simulation Interoperability Workshop*, 657–664 (1997).

[3] A. J. Gonzalez, R. F. DeMara, and M. Georgiopoulos, "Vehicle model generation and optimization for embedded simulation", *Proceedings of the Spring 1998 Simulation Interoperability Workshop*, 1385–1391 (1998).

[4] A. J. Gonzalez, M. Georgiopoulos, R. F. DeMara, and A. Henninger, "Automating the CGF model development and refinement process by observing expert behavior in a simulation", *Proceedings of the 7th Computer Generated Forces and Behavioral Representation Conference*, 251–256 (1998).

[5] A. Henninger, W. Gerber, R. DeMara, M. Georgiopoulos, and A. Gonzalez, "Behavior modeling framework for embedded simulation", *Proceedings of the 20th Interservice/Industry Training, Simulation and Education Conference*, 655–662 (1998).

[6] A. J. Gonzalez and R. Ahlers, "Context-based representation of intelligent behavior in training simulations", *Transactions of the Society for Computer Simulation* 15 (4), 153–166 (1998).

[7] W. J. Gerber and A. J. Gonzalez, "Real-time synchronization and modification of a behavioral vehicle model for embedded simulation", *Proceedings of the 21st Interservice/Industry Training, Simulation and Education Conference*, 859–868 (1999).

[8] P. Drewes and A. Gonzalez, "Instructor assistance using template based reasoning", *Proceedings of the IEEE International Conference on Systems, Man and Cybernetics, Volume 2*, 1918–1923 (1995).

[9] P. J. Drewes, *Automated student performance monitoring in training simulation.* Unpublished doctoral dissertation, University of Central Florida, Orlando, (1997).

[10] W. J. Gerber, *Parametric analysis of electrical load forecasting using artificial neural networks,* Unpublished master's thesis, University of Central Florida, Orlando, (1996).

A General Simulation Engine

William Foss

School of Electrical Engineering and Computer Science
University of Central Florida
Orlando, Florida 32816
Research Advisor: Dr. Mostafa Bassiouni

ABSTRACT

In the defense simulation community, we have seen the evolution of the Simulation Network (SIMNET) to Distributed Interactive Simulation (DIS) to High Level Architecture (HLA). Each step provided another level of interaction between existing and future simulations. As the use of simulation expands, the necessity to include variable levels of detail within complex real-time simulation exercises will increase. The discrete event simulation community has great experience in developing varied simulations models within the same simulation framework. The military training simulation community has experience developing complex physical and behavioral models that interact in real-time. This research presents some of the issues involved with combining the best practices and capabilities of these two simulation communities. This research also presents a small project that illustrates how a simple terrain representation and cognitive model can be modeled within an existing Discrete Event Simulation (DES) framework. This project uses the Silk simulation package and implements a small cognitive model of a driver interfacing with a physical vehicle model within a highway traffic environment. The hope is that this work will be extended to further unify the discrete event and military training simulation domains.

INTRODUCTION

The Department of Defense (DoD) has developed many major simulation systems for training, analysis, and testing. Each system has its own advantages and disadvantages. In an effort to build better and larger simulations, protocols, like SIMNET (SIMulation NETwork) and Distributed Interactive Simulation (DIS) [1], were developed to link similar systems together. As the protocols evolved, the DoD recognized the need for more than just a communications protocol, so they developed the High Level Architecture (HLA) [2].

The HLA works well in linking simulations designed for HLA, but there are significant overhead costs associated with HLA. First, the HLA is complex and takes some time to learn all the different requirements of the architecture. Second, converting an existing simulation system to use HLA requires a significant effort to support all the interfaces required by the HLA. Designing a new system requires similar effort to ensure that all the interfaces of HLA are supported. Finally, running an HLA simulation exercise requires one or more auxiliary programs to intelligently route traffic between simulation systems. These auxiliary programs add overhead to the execution of the simulation exercise. Some applications have been developed to ease the transition to HLA. The Institute for Simulation & Training developed an HLA Gateway [3] that allows existing SIMNET and DIS applications to interact with HLA. Even with a gateway, these applications will not support all the new features of the HLA.

These protocols and architectures have only addressed the linking of simulations together. When the DoD wants to develop a new simulation model, they currently have two choices: build a new simulation or add the model to an existing simulation. The first choice requires significant effort in recreating even the simplest parts of a simulation. The designers must create a network interface, an event manager, a terrain representation, a user interface, and many other parts. These parts of a simulation have been developed many times over in existing systems. Even though current designers can reuse much of the existing code, they still need to integrate and test the results. The second choice requires picking one of the many existing systems and designing a model specific to that system. The Army has made an effort to minimize the choices by developing requirements for OneSAF [4], which will be the single Semi-Automated Force (SAF) system for the Army. OneSAF is a proposed system that will integrate the best features of the various Army SAFs, like Modular Semi-Automated Forces (ModSAF) and Close Combat Tactical Trainer (CCTT) SAF, and provide a common architecture for future development of Army SAF models. One goal of OneSAF is to provide a common architecture for the other defense modeling and simulation (M&S) communities, while maintaining a primary focus on the needs of the Army.

Given the goal to create a common platform for all the Army's modeling and simulation (M&S) needs, the next step is to survey other simulation communities looking for features and capabilities that could address these issues with multiple resolutions, fidelities, and approaches to modeling.

One such area is the Discrete Event Simulation (DES) community. The DES community has developed several generic architectures that allow a modeler to create models in a very wide range of application domains. Zeigler developed a formal specification for designing simulation models called Discrete Event Systems Specification (DEVS) [5]. A number of commercially available packages allow modelers to use a Graphical User Interface to model and simulate a problem instead of developing code. Other packages have developed specific programming languages that add important capabilities to simplify model development.

OBJECTIVE

The objective of this research is to develop a simulation architecture that combines the complex distributed real-time aspects of the traditional military training simulation with the domain independence of a discrete event simulation package.

The short-term goal is to extend an existing system to support military training exercises and to support small discrete event simulation problems, like vehicular traffic flows. One approach to this first goal is to take an existing DES package like Arena [6] and modify it to support the complex physical, behavioral, and environmental models used in a distributed military training simulation. Another approach is to take an existing military training simulation and modify it to support simple generic discrete event simulations.

The long-term goal is to develop an open simulation package that would support the DES and the military training communities. One component of this effort would be to develop a model specification language that supports both simulation communities. This task may require a new specification language, or it could be as simple as extending an existing specification like DEVS.

SUPPORTING RESEARCH

Recognition of the Problem

A number of papers have investigated the issues involved with extending the current military simulation environment to support a more general, wide range of simulation models. The initial development of the High

Level Architecture specifically addressed some of these issues. Davis and Moeller [7] have illustrated that some problems exist with the HLA. Fujimoto [8] recognized the benefit to applying some of the parallel and distributed discrete event simulation techniques to the military training environments. Smith [8] expresses the need for "an architecture that supports an entire domain of simulation systems, providing a large pool of common functionality." He also mentions that standard architectures like HLA have helped, but more can be done to generalize and simplify the development of new simulation models.

Other Approaches

The unification of discrete event simulation and military training simulation has been the focus of several research efforts. The development of HLA made great strides toward integration of the analytical Aggregate Level Simulation Protocol (ALSP) [9] and the virtual training Distributed Interactive Simulation (DIS) environments. While both analytical and virtual training simulations could use the same infrastructure to communicate with other networked simulations, full and automatic interoperability between these simulations was not in the scope of the HLA. Many questions regarding interactions between multi-resolution models need to be answered before a virtual tank can fire upon an aggregate platoon and inflict realistic damage.

Earlier research performed at the Institute for Simulation & Training (IST) [10] illustrated that aggregate level constructive simulations and virtual level training simulations can be combined to increase training effectiveness both for commanders wanting more detail about a particular engagement and for training soldiers needing practice in the context of a large exercise. By linking aggregate level simulations like Eagle with virtual level simulations like ModSAF, this research discovered a number of fundamental issues that must be accounted for in a multi-resolution simulation [11]. While significant progress has been achieved by linking existing simulations, it is easier to address some of the more fundamental issues by implementing both levels of aggregate detail within a single simulation system [12].

TECHNICAL ISSUES TO SOLVE

The current and past research into discrete event and military training simulations has presented some approaches to address some interoperability problems between these simulation domains. This section identifies some of the larger research problems in the integration of discrete event simulations (DES) and military training simulation.

Real-time vs. Event timing

One of the most noticeable issues is the coordination of timing. While the "soft" real-time ordering of military simulation events should fit within the discrete event simulation (DES) environment, there are a number of continuous state changes that take place between events in the military training simulation models. Given a DES model that runs faster than real-time, it should be easy to slow the event processing to match with a real-time clock. However, DES models are not generally designed to run with any relation to the real-time clock. Some applications, like military doctrine analysis, want to run faster than real-time. Other applications like simulating a computer processor are acceptable to run much slower than real-time. This flexibility in timing hinders the wholesale adoption of DES models into a real-time environment. The complexity of the training simulation models also suggests that maintaining real-time processing of events may be difficult, especially when events start to cluster together.

Terrain Interoperability

The terrain and environment representation can vary widely in a discrete event simulation, while the military training simulation has traditionally used polygonal databases, as found in the Compact Terrain DataBase (CTDB) format [13], to coarsely represent the terrain skin, buildings, trees and other artifacts in the environment. A common approach to solving terrain interoperability issues between simulation systems at the same level of resolution and fidelity is to use the same terrain format. Another approach is to translate to between formats while ensuring a high level of correlation. More correlation problems arise when the terrain resolutions differ by several orders of magnitude. Consider a model of a radar system that only needs to detect major roads, buildings or terrain features operating in the same exercise where an individual combatant is entering a house and needs to determine if there is an enemy behind the couch in the corner of the living room. With too much detail the radar model will not be able to process the information within the computation time available, but with too little detail, the soldier entering the house will not have proper representations of the couch.

Complex Behavioral Models

One of the traditional strengths of the military training simulations is the high level of detail and accuracy designed into the simulation models. The models have strived for as accurate representations as the current computation power can handle within real-time constraints. While DES systems can often provide more detailed representations while sacrificing the real-time requirement, many of the military DES models in the military analysis communities have focused on much faster than real-time responses to provide quick feedback for mission planning.

Merging the capabilities of DES and military training simulations would require a component structure for plugging in different simulation models depending on the desired size, speed, resolution, and fidelity of the simulation exercise. Ideally, if a simulation exercise needs to have the context of 10,000 entities at the vehicle level with real-time interaction with training soldiers, then the fidelity of the vehicle models could be scaled to fit the available computational resources. In another scenario, the real-time requirement could be relaxed to allow a very detailed exercise with 10,000 entities modeled using high fidelity vehicle level models.

CURRENT RESEARCH

The research project was a simple investigation into the possibility of creating more complex training like behaviors within a discrete event simulation environment. This problem was approached by taking an existing commercially available DES package, Silk, and developing a simple behavior model based on cognitive models used in Soar [14], ACT-R ("Atomic Components of Thought" Research) [15], and COGnition as a NEtwork of Tasks (COGNET) [16].

Goals

This research project provides a demonstration of a simple cognitive model implemented within the framework of a discrete event simulation package. It is designed to support further investigation and extension into the application of cognitive modeling within discrete event simulation. The object-oriented architecture of this system allows many of the objects to be reused and extended in future simulation projects.

The first subsection documents the design of an object-oriented simulation based on the Silk simulation package. This documentation includes the use cases, requirements, and design diagrams represented in the Universal Modeling Language (UML) [17]. The next subsection documents the lessons learned during the development process. Finally, the last subsection discusses some opportunities for extension and further development in other projects.

Background

This section provides background on three major components of the project. The Silk simulation package provided the infrastructure for the discrete event simulation. The cognitive modeling community provided the basis for the behavioral model developed in this project. The traffic simulation community provided an application domain to create an example simulation.

Silk

This project uses Silk® [18] for the management of entity-based events in the simulation. Silk is a package of Java classes developed by ThreadTec, Inc. to support development of discrete-event simulations. It is designed simplify development of entity and process based models within an object-oriented simulation environment. The choice of a Java-based environment allows the simulation models to run on a wide range of hardware and software platforms. More information about Silk can be found at the ThreadTec web site. [18]

Cognitive Modeling

Some recent developments in the military simulation domain have made cognitive modeling a major focus of the development for new simulation systems. The current training simulations focus on doctrine to provide realistic behaviors for Computer Generated Forces (CGF) [19]. Strict adherence to doctrine has exposed some limitations when the current situation does not match exactly with one of the situations identified in doctrine. The model also has problems when extra information in the scenario would tell a real human that doctrine is not appropriate, but the CGF model does not recognize these clues. The realism of CGF behaviors is an important factor in training, because often the CGF simulates the opposing forces (OPFOR) behavior. If the OPFOR do not exhibit realistic behavior, then the soldiers in training may not be prepared for the real OPFOR behaviors found on the battlefield.

Cognitive modeling is also used in many other domains. Human factors researchers use cognitive models to simulate human performance in user interfaces. This information is used to improve the components and layout of user interfaces. Several applications, like COGNET, Soar, and ACT-R, aid in the development of cognitive models.

Traffic Simulation

This research used the traffic simulation community as an application domain. This project was able to quickly develop a simple traffic simulation to demonstrate linkage between the cognitive model and Silk environments. The author also has a personal interest in traffic simulation, because he spends about two hours each workday driving across Orlando.

Design

This section presents the design of the objects and models used in this simulation project. It is divided into three subsections that will describe the use cases, the cognitive model, and the situational awareness model. The

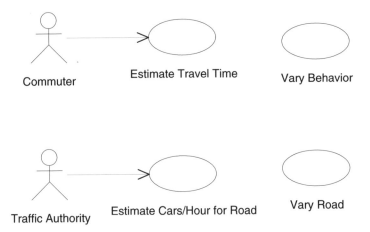

Commuter   Estimate Travel Time   Vary Behavior

Traffic Authority   Estimate Cars/Hour for Road   Vary Road

Figure 1. Cognitive Driver Simulator Use Cases.

diagrams illustrated below for these models were developed in UML notation using Rational Rose 4.0 Student Edition.

Use Cases
    This project has two primary use cases as illustrated in Figure 1. The
first use case is the evaluation of traffic conditions by a commuter. The
commuter may look at scenarios where traffic congestion is higher based
on the time or direction of travel. The second use case allows a traffic engineer to simulate and predict traffic flows based on the number of cars and
number of lanes available.
    Some of the potential uses for this type of simulation are:

• Determining statistical flow of traffic from point x to point y.
• Estimating average commute time from x to y under certain traffic conditions,
• Creating statistically accurate congestion vehicles.

Cognitive Model
    The focus of the cognitive model is to demonstrate the use of a cognitive
model within a discrete event simulation environment. The components of
the model were modeled after other cognitive modeling architectures. These
components are represented in Figure 2.
    Objects in the Cognitive Model are:

• Perception—the sensors poll the instruments to determine if any new
information can be gathered. The new information is placed in working
memory for later use.

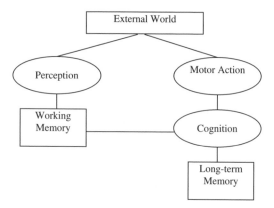

Figure 2. Cognitive Driver Model.

- Cognition—the cognition polls working memory to determine if new information has been sensed. Once new information is received, the cognition determines if an action is required. If so, the cognition directs the motor action to perform the necessary action. This object will be actively looking for information to process.
- Motor Action—the motor action provides an interface for the cognition object to implement the desired actions. It passively accepts commands from the cognition object.
- Working Memory—the working memory provides a storage place for all the situational awareness information. This component passively accepts queries and updates from the working memory. It also accepts updates from the perception component.
- Long-term Memory—the long-term memory is currently defined in the simulation code, but this structure could be placed in a data file to allow better configuration of long-term memory. A future design could allow for this memory to be encoded in a simple modeling language for easy configuration.
- External World—the model of the external world is designed to clearly separate the physical and behavioral models. In this way, the system models the interface between a human and the vehicle. The model could interface to more detailed models of the vehicle components, but they are not vital to this simulation, so they have been abstracted into two attributes, speed and location. More detail about the physical model is included in the next section on implementation.

The interface between the cognitive model and the human vehicle interface is designed for future training environments where a human would control the interface to the vehicle. This adaptation would require transition to a real-time event simulation and would require development of a simple Graphical User Interface (GUsI) to the methods controlling speed and direction. This research could allow a future comparison of real human behavior with cognitive model behavior.

Situational Awareness

The driver's terrain environment is represented by a 6x3 matrix in which the driver is positioned in the 2nd column and 4th row. This matrix represents the area of interest around the vehicle. Each cell may contain one of three value types. The first value represents clear road. The second value represents no road available, for example, when the driver is in the right most lane there is no road to the right. The final value represents an occupied cell. In this case, the value of the cell provides a reference to the vehicle occupying it.

The matrix will be indexed as follows:

	Left (L)	Center (C)	Right (R)
3	3L	3C	3R
2	2L	2C	2R
1	1L	1C	1R
0	0L	0C	0R
-1	-1L	-1C	-1R
-2	-2L	-2C	-2R

Implementation

The implementation of this simulation follows a recent idea to separate the physical model from the behavioral model. By separating these components, we allow the behavioral model to be reused to control several objects with a similar physical model. The original physical model can also be tested with several different behavioral models allowing the user to compare the effectiveness of each. This section is composed of four parts. The first part covers the general development issues. The second part covers the physical model implementation. The third part covers the cognitive model implementation. The final part covers the environmental model implementation.

General

Based on the existing capabilities of the Silk package, this project used a single event queue structure designed to run on a single processor. This design process began by identifying the concepts and intended use of the simu-

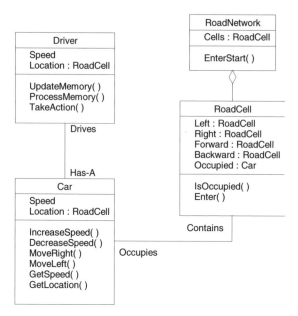

Figure 3. Cognitive Driver Simulator Class Diagrams.

lation software. These designs were modeled using several methods including UML when appropriate. Figure 3 diagrams the main classes and connections in the component design. As the designs were refined, object-oriented coding techniques were mapped to the design for easy implementation.

Physical Model

The physical model in this simulation represents the physical capabilities and properties of the vehicle. Initially, the car object only modeled two attributes, but it could be extended to include the acceleration, braking distance, weight, width, drag, and fuel capacity parameters. The first attribute represents the speed of the car. The second attribute identifies the location of the car in the environment. The location information provides a link to the road cell (see Environment Model below) so that the car can identify objects in the surrounding cells. These attributes are read by the UpdateMemory method of the Driver Object. This method represents the sensor component of the cognitive model. The GetSpeed and GetLocation methods in the Car object provide access to these attributes. The car model also represents the physical limitations of certain parameters like speed.

The car object also provides several methods that are used by the driver object to manipulate the physical model. These methods are IncreaseSpeed, DecreaseSpeed, MoveLeft, and MoveRight. These methods provide the interface to the motor action component of the cognitive model. The physical

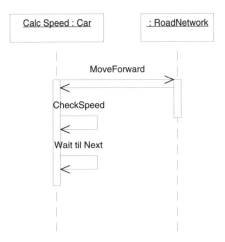

Figure 4. Cognitive Driver Simulator Car Process Diagram.

model also represents the continual motion of the car as diagramed in Figure 4. The car will schedule events to move forward based on the current speed and estimated time to get to the next cell.

Behavioral Model

The driver object provides the cognitive behavior aspect of the simulation. The components of the cognitive model are represented three tasks, as diagramed in Figure 5.

The first task updates the working memory from the car sensors. These sensors include inputs of destination, surrounding vehicles, and lane information. This task is performed in the UpdateMemory method. This method gets the current speed of the car and the location of the car within the environment.

The second task searches the rules to determine if the update in the memory triggers any actions to take place. This task is performed by the ProcessMemory method. This method uses three simple rules.

- If there exists a car in cell 1 spaces ahead whose speed is less than 5 mph greater than mine, then slow down to avoid collision.
- If there exists a car in cell 2 spaces ahead whose speed is less than 10 mph greater than mine, then slow down to avoid collision.
- If there exists a car in cell 3 spaces ahead whose speed is less than 15 mph greater than mine, then slow down to avoid collision.

Further rules can easily be added, but these rules were sufficient for the scope of this project.

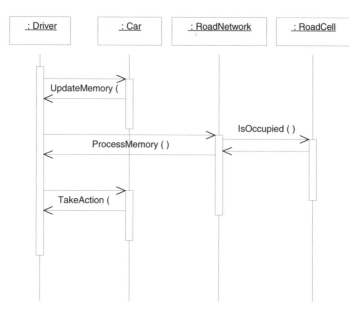

Figure 5. Cognitive Driver Simulator Driver Process Diagram.

The third task uses the motor action interfaces to modify the physical model in the external world. This task is performed by the TakeAction method.

Environment Model

The environment model consists of a set of road cells, which are connected into a road network. Figure 6 illustrates the structure of the terrain model. The road network is a two dimensional array of road cells. Each cell contains links to the four possible cells that can be reached from that cell. The cell also contains an indication of the intended direction of travel within the cell. Cells in the left most lane do not link to cells to the left. Likewise, cells in the right most lane do not link to cells to the right. At an exit ramp, the cells continuing with the road and the exit ramp cells cease to have links between them, once the ramp is considered separate from the highway. The vehicle can exit the simulation by moving off the known road. Each cell also has a link to a car (if present) that occupies that cell. Each cell in the network is one lane wide and a quarter mile long. The road network implemented is a ten mile segment of a three-lane highway.

Experimental Results

The project development presented several lessons learned about the design and development of object-oriented simulation. The development

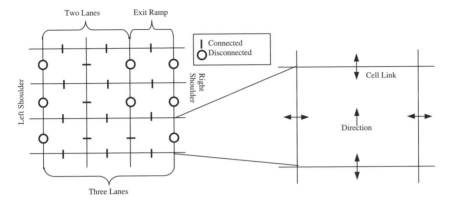

Figure 6. Simple Terrain Structure.

of this project illustrated the need to diagram and revise. The first few drafts of class diagrams did not provide the necessary capabilities. After a few revisions, both before and during coding, the final revision was established.

This project has provided some insight in the development of a cognitive model within an object-oriented discrete event simulation. While this implementation is limited in several respects, it does demonstrate an object-oriented implementation of a cognitive model within a discrete event simulation. This paper has documented some of the important aspects of design and implementation with cognitive models and discrete event simulation.

CONCLUSIONS

In the defense simulation community, we have seen the evolution of SIMNET to DIS to HLA. Each step provided another level of interaction between existing and future simulations. As the use of simulation expands, the necessity to include variable levels of detail within complex real-time simulation exercises will increase. The discrete event simulation community has great experience in developing varied simulations models within the same simulation framework. The military training simulation community has experience developing complex physical and behavioral models that interact in real-time. This research has identified some of the issues involved with combining the best practices and capabilities of these two simulation communities. This research also presented a small project that illustrates how a simple terrain representation and cognitive model and be modeled within an existing DES framework. The hope is that this work will be extended to further unify the discrete event and military training simulation domains.

FUTURE WORK

Obviously, there is significant work left in the area of unifying the military training and discrete event simulation domains. The first extension of this research will be further investigation of existing simulation architectures, like DEVS, Silk, and HLA, to determine the potential benefits of extending these architectures to better support both communities. Zeigler has done some work in integrating the HLA and DEVS environments, but further research is needed to see how well the DEVS/HLA system addresses the issues with terrain interoperability and complex real-time models. Another area of research is to investigate the ability of existing military training simulations, like ModSAF, to interoperate with DES models through the HLA. Other approaches to investigate include exploring the current component based software design and development architectures, like CORBA and Enterprise Java Beans, to see if they provide better support for a more generic simulation and modeling architecture.

ACKNOWLEDGEMENTS

I would like to gratefully thank the Link Foundation for their support of this research. I would also like to thank the Institute for Simulation and Training and the School of EECS at the University of Central Florida for their support of my continuing education. Finally, I would like to thank Jesus Christ for dying on the cross and being raised from the dead, so that I might be forgiven my sins and spend eternity with him.

REFERENCES

[1] Standards Committee on Interactive Simulations (SCIS), *IEEE Standard for Distributed Interactive Simulation—Application Protocols (IEEE Std 1278.1–1995)*, IEEE Standards Board (September 21, 1995).
[2] "DMSO: High Level Architecture" available at *http://hla.dmso.mil/index.php? page=64* Defense Modeling and Simulation Office (last accessed Oct 2000).
[3] D. Wood and M. Petty, "HLA Gateway 1999" *Proceedings of the Spring '99— Simulation Interoperability Workshop,* Orlando, FL (1999).
[4] *One Semi-Automated Forces Operational Requirements Definition* (November 1997).
[5] B. Zeigler, *Theory of Modeling and Simulation.* John-Wiley, New York (1976).

[6]   "Arena: The Preeminent Simulation Solution" available at *http://www.sm.com/ sim/arena/default.htm* (last accessed Oct 2000).

[7]   W. Davis and G. Moeller, "The High Level Architecture: Is There a Better Way", *Proceedings of the 1999 Winter Simulation Conference* (1999).

[8]   E. Page (Moderator), et. al. "Panel: Strategic Directions in Simulation Research", *Proceedings of the 1999 Winter Simulation Conference* (1999).

[9]   "Aggregate Level Simulation Protocol (ALSP)" available at *http://ms.ie.org/alsp* (last accessed Oct 2000).

[10] D. Stober, "Survey of Constructive + Virtual Linkages", *Proceedings of the Fifth Conference on Computer Generated Forces and Behavioral Representation (IST-TR-95–04)*, Orlando, FL, pp. 93–102 (1995).

[11] S. Schricker, R. Franceschini, D. Stober, and J. Nida, "An Architecture for Linking Aggregate and Virtual Simulations" *Proceedings of the Sixth Conference on Computer Generated Forces and Behavioral Representation.* Orlando, Florida, pp. 427–434 (1996).

[12] R. Franceschini, "A Simple Multiple Resolution Entity Simulation" *Proceedings of the Eighth Conference on Computer Generated Forces and Behavioral Representation.* Orlando, Florida, pp. 597–608 (1999).

[13] J. Smith, "Recent Developments in ModSAF Terrain Representation." *Proceedings of the Fifth Conference on Computer Generated Forces and Behavioral Representation.* Orlando, Florida, pp. 375–381 (1995).

[14] The Soar Group, "Soar Home Page", available at http://bigfoot.eecs.umich.edu/~soar/ University of Michigan (last accessed Oct 2000).

[15] "ACT Research Home Page", available at http://act.psy.cmu.edu/ACT/act/actr.html Carnegie Mellon University (last accessed Oct 2000).

[16] "COGNET—COGnition as a NEtwork of Tasks" available at http://www.chiinc.com/cognethome.shtml CHI Systems, Inc. (last accessed Oct 2000).

[17] "Introduction and Getting Acquainted with UML" available at *http:// www.rational.com/uml/gstart/index.jsp* Rational Software Corporation (last accessed Oct 2000).

[18] "What is Silk?" available at http://www.threadtec.com/silkintro.html ThreadTec, Inc. (last accessed Oct 2000).

[19] R. Pew and A. Mavor: *Modeling Human and Organizational Behavior: Application to Military Simulations*, National Academy Press, Washington, D.C. (1998).

PART III

OCEAN ENGINEERING AND INSTRUMENTATION

Initial Water Entry at Oblique Angles

Carolyn Q. Judge

Department of Naval Architecture and Marine Engineering
University of Michigan, Ann Arbor, MI 48109
Research Advisor: Armin Proesch

ABSTRACT

The water impact of planing boats operating at high speed is known to cause significant discomfort, loss of control of the vessel, occasional passenger injury and possible capsize. How a planing hull reacts when impacting the water is very important to high performance boat designers. The method described here is a two-dimensional impact model that allows for asymmetric vessel geometry and horizontal impact velocity. Two types of impact flow are established based on the degree of asymmetry and the ratio of horizontal to vertical impact velocity. Type A impact occurs when the flow stays attached to the hull until it reaches the chines. For Type B impact, the flow on one side of the hull separates from the hull at the keel. A two-dimensional vortex distributions is applied for solving the nonlinear boundary value problem. The initial conditions are determined from the basic solutions for straight-sided contours with constant impact velocities. Experimental drop tests of a prismatic wedge were performed to gain understanding of initial water impact when asymmetry and horizontal impact velocity are involved. In particular, the initiation of Type B flow was investigated. Experimental investigation into initial impact for both Type A and Type B flows is presented and compared with predictions from this method.

INTRODUCTION

High-speed planing boats are widely popular but little is understood about their stability at high speeds. Many of these craft are known to experience unexpected behavior at operational speeds. Research at the University of Michigan intending to understand dynamic instability has used a water impact model to determine the flow over a cross-section of the hull. The impact model takes a two-dimensional section of the hull and predicts how the flow moves over the bottom as the hull section enters the water. By using a low order strip theory and viewing the planing hull as a series of cross-sections at different points of impact (near the bow the hull is just starting to enter the water while near the transom the hull has mostly entered the water), this model determines the transverse flow characteristics over the entire hull. The resulting boundary value problem can be numerically solved using a two-dimensional vortex distribution.

In this paper a model is established for asymmetric vessel impact flows with horizontal impact velocity. Initially separated flow due to asymmetry and horizontal impact velocity is investigated. Initial conditions are determined from the basic solutions for straight-bottomed hull sections (with constant deadrise) during constant velocity impact. The numerical solution process involves discretization of the hull and an iterative solution technique. The method of two-dimensional vortex distributions is employed to model the boundary value problem. Experimental investigation into initial flow separation off the keel during impacts including asymmetry and horizontal impact velocity is described. Comparisons between the present model and experimental results are presented.

PROBLEM DESCRIPTION

Background

Water impact analysis was first studied because of interest in seaplane landings. In 1929 von Karman provided the first theoretical solution for determining the bottom pressures on the hull of a landing seaplane [1]. Wagner, in 1932 [2], then used an expanding flat plate model to calculate the local water surface elevation. Recently, Vorus [3] developed a flat cylinder model for vertical two-dimensional impact flows and Xu et. al. [4] extended this theory to include geometrically asymmetric impact. This theory exploits the fact that as the section contour becomes "flatter" the transverse flow perturbation velocity tends toward infinity. As the cylinder flattens toward the horizontal axis, the boundary conditions on the axis apply with increasing accuracy, implying a limit of geometric linearity. However, the increase in transverse flow perturbation velocity resulting from this flatness implies increasing hydrodynamic nonlinearity. This method is geometrically linear, that is the flat cylinder boundary conditions are applied on the horizontal axis, but it is hydrodynamically nonlinear and so it fully retains the large perturbation flow produced by the impacting flat cylinder in the axis boundary conditions.

As has been done traditionally, assumptions of zero gravity, incompressibility and zero viscosity are used to simplify the planing or water impact model.

A body-fixed coordinate system is used with y and z representing the horizontal and vertical axes, respectively. A small elevation angle, $\beta(y)$, is assumed. The vertical downward velocity of the cylinder, $W(t)$, is prescribed and $Z_{WL}(t)$ represents the height of the undisturbed water surface above the lowest body coordinate, the apex (point O), or keel.

On impact, the free surface is turned under the body section and forms an initially attached jet, as shown in Figure 1. The points B_1 and B_2 are called the jet "spray-roots" and advance rapidly outward along the cylinder contour. B_1 and B_2 are closely followed by C_1 and C_2, the "zero-pressure" points. For y-coordinates greater than C_1 and less than C_2 the dynamic pressure is zero. While B_1 and B_2 are attached to the contour, the contour pressure has a sharp spike with a large negative gradient into C_1 and C_2. The large pressure gradient is a result of the large flow accelerations near C_1 and C_2. As time advances, both the zero-pressure points and the jet spray-roots move outward.

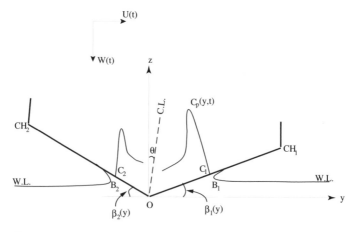

Figure 1. Type A model of cylinder impact (small asymmetry) and horizontal impact velocity.

Xu *et. al.* [4] defined two types of asymmetric impact. Type A flow is when there is small asymmetry and the flow moves outward along the contour on both sides of the apex. Type B flow occurs when there is large asymmetry and the flow separates from the contour at the apex on one side.

Without loss of generality, $\beta_1 \le \beta_2$ is assumed near the apex. The angle of heel represents the asymmetry and is defined by

$$\theta = \frac{\beta_2(0^-) - \beta_1(0^+)}{2} \quad , \tag{1}$$

where $\beta_1(0^+)$ and $\beta_2(0^-)$ are the elevation angles at the apex on the right and left sides, respectively. The zero-pressure points, C_1 and C_2, are also called the wetted points, between which the wetted surface, with non-zero dynamic pressure, is defined. C_p is the hydrodynamic pressure coefficient, W(t) is the vertically downward velocity, and U(t) is the horizontal velocity to the right. The body-fixed coordinate system is shown by O_{yz} in Figure 1 where y and z represent the horizontal and vertical axes, respectively.

The elevation angle β_1 is assumed to be small and characterizes the bottom flatness. It is appropriate to consider that Type A flow always attaches to the bottom before moving out along the contour. The onset of Type B flow is calculated from the initial positions of the zero-pressure points (C_1 and C_2) for different θ. As θ increases, C_2 moves closer towards the apex. When C_2 reaches the apex, the jet flow on the left side is forced to separate. The impact flow of Type A then turns into Type B impact flow. The Type A model includes an infinite velocity at the sharp apex, which generally exists for any asymmetric flow.

Xu's asymmetric model, like Vorus's model, considers only vertical impact, i.e. the horizontal velocity component of impact, U(t), is zero. In addition to the vertical impact velocity and possible asymmetry, the present model allows for the body to travel in the horizontal direction before and during impact. A symmetric body impacting with horizontal velocity will produce Type A and Type B flows, when roll is constrained. The horizontal component of the impact velocity, U(t), causes the flow to move along the contour faster on one side then on the other, creating hydrodynamically asymmetric impact.

Solution Method

The total normal and tangential contour velocities in terms of the vertical and horizontal velocities are

$$\overline{V}_n = (W + w)cos\ \beta + (U—u)sin\ \beta \ , \tag{2}$$
$$\overline{V}_s = (W + w)sin\ \beta—(U—u)cos\ \beta \ , \tag{3}$$

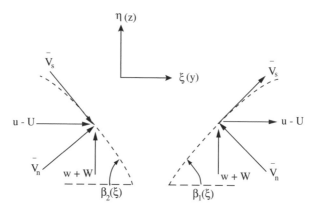

Figure 2. Contour Velocity.

where W and U are the free stream, or "fall", velocities and w and u are the perturbation velocities. Due to the different elevation angles on the two sides,

$$\beta = \begin{cases} \beta_1(y) \ if \ y > 0 \\ -\beta_2(y) \ if \ y < 0 . \end{cases} \tag{4}$$

Figure 2 gives a vector representation of the body boundary conditions. Rolling velocity due to asymmetric impact is considered of higher order and is neglected.

A vortex distribution is used to solve the resulting boundary value problem. Bound vortices are placed on the wetted body surface, between C_1 and C_1. Free vortices are shed at these points and form jet flows that follow the body contour. A condition of zero tangential velocity outside the jet spray-roots (B_1 and B_2) identically satisfies the condition of atmospheric pressure on the free surface. The flatness due to small elevation angle is exploited by collapsing the body and free surface contours to a horizontal level for satisfying the boundary conditions. The vortices are arranged on the contour such that they satisfy the boundary conditions. The strength of the vortex distribution is

$$\gamma(\xi,t) = -2u(\xi,t) , \tag{5}$$

where u is the perturbation velocity in the horizontal direction.

Initial Wetting Conditions

The body contour is assumed to have constant, nonzero elevation angles, β_1 and β_2, near the apex. In addition, in the small time after initial impact the velocities of the impact can be taken as constant at the initial values, $W(0) \equiv W_0$ and $U(0) \equiv U_0$. The locations of the zero-pressure points and jet spray-roots as well as the jet velocities are determined using the boundary condition equations and an iterative method.

Initiation of Type B Flow

Following Xu's [4] investigation of asymmetry in vertical impacts, the limits of asymmetric impact with horizontal impact velocity are examined, in particular the transition between Type A and Type B impact. As the asymmetry, θ, increases, the zero-pressure point on the left-hand side, C_2, moves back toward the apex. When C_2 reaches the apex, the impact becomes Type B. The phrases

Type B impact and ventilation are used interchangeably when discussing this limiting behavior. Figure 3 shows the limiting angle of β_2 versus the corresponding β_1 at which Type B flow is initiated.

Figure 4 shows the critical value of U/W versus the corresponding elevation angles at which ventilation occurs off the apex. For symmetric bodies, i.e. $\theta=0$, and small elevation angles, the horizontal velocity must be much greater than the vertical velocity for ventilation to occur. However, as the elevation angle increases the required ratio decreases rapidly and then flattens out. Thus, the critical value of U/W is less for bodies of larger elevation angles.

For a given ratio of horizontal to vertical impact velocity, the angle β_2 required for ventilation can be determined. In Figure 3 the limiting angle of β_2 versus the corresponding β_1 is shown for different ratios of impact velocities. For each of the different ratios, the dependence of β_2 on β_1 for the

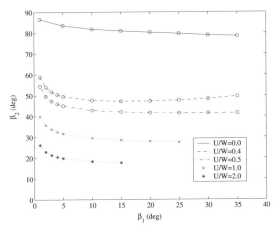

Figure 3. The critical angle, β_2 versus the corresponding β_1 at which ventilation occurs off the apex for different ratios of impact velocities.

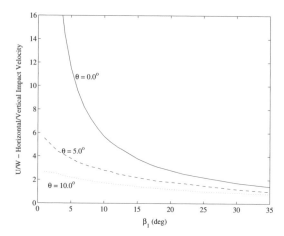

Figure 4. The critical value of U/W versus the corresponding β_1 at which point ventilation occurs off the apex for different angles of θ.

initiation of Type B flow is small. An increase in horizontal velocity decreases the value of β_2 required for ventilation. Figure 4 shows the critical value of U/W versus β_1 for different degrees of asymmetry (as measured by θ). As expected, larger degrees of asymmetry reduce the critical ratio of impact velocities.

EXPERIMENTAL INVESTIGATION OF INITIAL WATER IMPACT

Experimental Set-Up

Drop tests including asymmetry and horizontal impact velocity were carried out at the Marine Hydrodynamic Laboratory at the University of Michigan, Ann Arbor. The objective was to investigate the fluid motion during initial water impact of a prismatic body. Investigation of the behavior of flow ventilation from the body as well as the onset of Type B impact flow were also motivations for the experimental study. The set-up consisted of a slide assembly with guide rails and linear bearings, a cart that ran on the linear-bearing slide assembly, a metal arm connected to the cart, a symmetric prismatic wedge attached to the metal arm, and a free-falling weight that pulled the cart along the linear-bearing slide assembly. The weight was attached to the cart by a wire and a system of pulleys that allowed the weight to fall vertically after the cart was released. The wedge ran parallel to the linear-bearing slide assembly but offset a small distance. The wedge could be rotated to different heel angles and was restrained from roll dur-

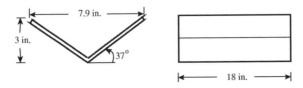

Figure 5. Schematic of Drop Model (wedge).

ing impact. The linear-bearing slide assembly could be set at varying angles to produce different ratios of horizontal to vertical impact velocities. The cart was released from the top of the linear-bearing slide assembly by the release of an electric magnet and was pulled down the linear-bearing slide assembly by the free-falling weight. The prismatic wedge was made of aluminum and had a length of 18 inches, a width of 7.9 inches and an elevation angle of 37°. A schematic of the test model is shown in Figure 5.

The impact was recorded using a Kodak EktaPro high speed camera with a 100mm lens set at a frame rate of 2000 frames per second. A Liconix argon ion laser produced a light sheet perpendicular to the apex line and approximately 5 inches from the edge of the body. The laser illuminated the water and wedge during the initial water impact. Fluorescein dye and particles, which consisted of hollow glass spheres (110P8) coated with 2–6018 adhesive, were used to help visualize the impact flow and to detect the wetted surface. The velocity of the body was measured using sequential video frames. The wedge was dropped into initially quiescent water. At least three releases were done for every test configuration. Figures 8, 10, 11, and 12 are typical examples of the video recorded drops.

Two main observations were made of the fluid motion during the impact. The first was measuring the ratio of wetted points for different ratios of impact velocities and degrees of asymmetry. The second was determining the parameters for onset of Type B flow and investigating the behavior of the fluid ventilation. Figures 6, 7, and 9 compare the experimental wetted point ratios with the predicted ratios. Figure 12 shows an example of flow ventilation.

Experimental/Numerical Comparison

Table I gives the different test configurations, i.e. the ratios of impact velocities and the degrees of asymmetry that were used for the drops. The video was digitized and the data taken off the digitized pictures. The measurements made from the images were the horizontal distances between the apex and the wetted points on each side of the body. The wetted point is defined to be the location where the curved surface of the water inter-

Table I. Experimental Test Configurations

U/W	θ $(\pm 2°)$
0.0	$0°, 5°, 10°, 15°, 20°, 25°, 30°, 34°^\dagger$
0.25	$0°, 5°, 10°, 15°, 20°, 25°, 30°, 34°^\dagger$
0.5	$0°, 5°, 10°, 15°, 20°, 25°, 30°, 34°^\dagger$
0.75	$0°, 5°, 10°, 15°, 20°, 25°, 30°, 34°^\dagger$
1.00	$0°, 5°, 10°, 15°, 20°, 25°, 30°, 34°^\dagger$
1.33	$0°, 5°, 10°, 15°, 20°$
2.00	$0°, 5°, 34°^\dagger$

†*Note:* The maximum θ was difficult to maintain during the drops, so although 34° was attempted, this category is better viewed as θ_{MAX}.

sects the body and is expected to compare approximately to points B_1 and B_2. For each drop, the first three or four frames were used to measure the wetted points and the measured ratios of B_2 to B_1 were then averaged.

In the case of a purely geometric impact (no hydrodynamic effects included), the location of the wetted points is simply at the intersection of the undisturbed free surface and the body. This can be determined entirely by geometry. Let the distance from the apex to the intersection of the undisturbed free surface on side 1 (the right side) be called X_1 and the same distance on side 2 (the left side) be called X_2. The ratio of these two points is defined as d. For a symmetric body, d is equal to 1. As the heel angle, θ, increases, d becomes less until θ equals the elevation angle β_1 at which point, $d=0$. Figure 6a shows how d varies with θ for a symmetric wedge with an elevation angle of 37°. This graph is unaffected by adding horizontal velocity since the change in d is determined solely by geometry.

Hydrodynamic effects during impact create jets that move along the body on either side of the apex. These jets are created where the free surface intersects and moves quickly along the body, creating a curved free surface. In other words, the intersection of the water surface and the body is not at the geometric position of the intersection of the undisturbed free surface and the body surface. The distances from the apex to these wetted points (which are due to both geometric and hydrodynamic effects) are approximated as B_1 and B_2 on the right and left sides, respectively. The ratio of these points is defined as c_2^*. Figure 6b shows a graph of c_2^* versus θ where the ratio of impact velocities is $U/W = 0.25$. In this case, the addi-

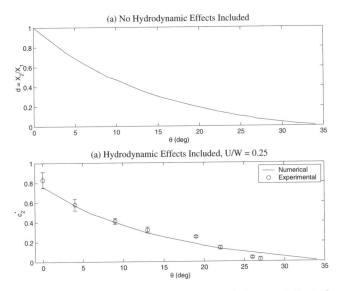

Figure 6. Graph of the ratio of X_2 to X_1 for asymmetric impact (effect of geometry, no hydrodynamics). (b). Graph of the ratio of B_2 to B_1 for impact with a ratio of impact velocities of $U/W = 0.25$ (hydrodynamic effects included).

tion of horizontal velocity to the impact does effect c_2^* since the hydrodynamic effects are included.

The importance of the hydrodynamic effects varies depending on the type of impact. For some impacts c_2^* is very similar to d, while for other impacts, the two ratios are very different. If B_2 and B_1 are proportional to X_2 and X_1, then, despite the fact that the physical position of the points are not the same, d and c_2^* would be equal. If the hydrodynamic effects are dominant, then the positions would not be proportional and d and c_2^* would be very different.

Graphs of the difference between d and c_2^* for different amounts of asymmetry and horizontal impact velocity are shown in Figure 7. When there is no horizontal velocity, simply vertical impact with asymmetry (Figure 7a), there are only small differences between the predictions for c_2^*, the experimental results for c_2^*, and the geometric ratio d. The numerical solution captures the slight curve in the experimental data, but in this type of impact the geometric asymmetry dominates the flow and the prediction based on pure geometry (the intersection of the body with the undisturbed free surface) gives fairly good results.

When horizontal impact velocity is present, however, even when the geometry is symmetric, the flow is asymmetric. This asymmetry is dominated by the hydrodynamics and diverges from the purely geometric pre-

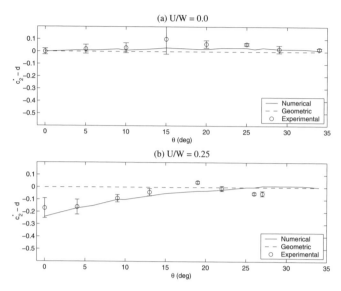

Figure 7. The difference between predicted c_2^* and d for two methods and for experimental data.

diction. When θ is large, the geometric asymmetry dominates the flow and c_2^* is very close to d. When θ is small and there is horizontal impact velocity the hydrodynamics become very important and c_2^* diverges from d.

Typical video footage of Type A impact is shown in Figure 8. The body contour lines are added to the actual video, but the curved free surface lines are due to the reflection of the particles in the water. The frames are sequential ($\Delta t = 0.5$ msec) and it is clear how the jets move out along the body as the impact progresses.

Investigation of Initial Flow Ventilation (Type B Impact)

Figure 9 shows c_2^* versus θ for an impact with ratios of 0.5 and 0.75 for horizontal to vertical impact velocity. This configuration was of interest because ventilation did not occur for the symmetric case, but did occur before the maximum θ was reached. In Figure 9 ventilation is indicated by c_2^* becoming zero. Table II compares the theoretical predictions for ventilation with the ventilation limits measured experimentally. The change in experimental is 5° increments. Table II is to the nearest 5°. Overall, the theory does quite well.

To study the transition from Type A to Type B flow, video was taken of drops at increasing θ for a given velocity ratio. Figure 10 shows pictures taken at approximately the same point of impact for different θ's at a veloc-

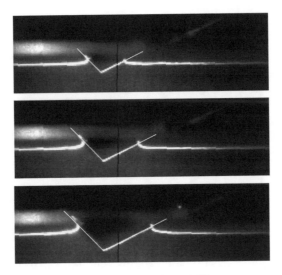

Figure 8. Type A Impact with $\theta = 10°$ and U/W = 0.25.

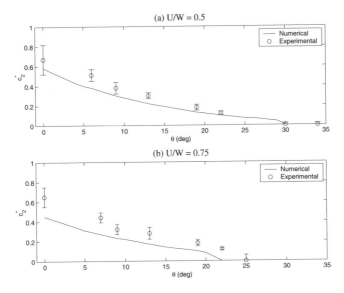

Figure 9. The ratio of B_2 to B_1, i.e. c_2^*, versus θ for (a) U/W = 0.5 and (b) U/W = 0.75.

ity ratio of U/W = 0.75. For θ=15° the flow is smooth along the body and it exhibits similarity flow type behavior, there is no ventilation. For a θ of 20° the flow is still attached, but it is beginning to flare out away from the apex. The jet is still attached to the body, however, so this is still a Type A impact. For θ greater than 20° the flow is clearly separated from the body at the jet

Table II. Experimental and Theoretical Predictions for Initiation of Ventilation

U/W	θ_{SEP} (Experimental, $\Delta\theta_{SEP} = 5°$)	θ_{SEP} (Theoretical)
0.5	30°	30°
0.75	25°	20°
1.0	15°	10°
1.33	5°	5°
2.0	0°	0°

location as well as moving off the body at the apex. Increasing the asymmetry makes these effects more obvious. Defining similarity type flow to mean that the flow looks the same at different points in time, the transition region as the body first impacts the water, but before ventilation has occurred, is not similarity type flow. Some of the video records show that for Type B flow, the shape of the free surface evolves with time. Therefore, Type B flow cannot be assumed to be similarity type flow even after the flow has detached from the body.

The behavior of the flow as the wedge impacts the water depends on the degree of asymmetry and the amount of horizontal impact velocity. When ventilation is just beginning the flow starts initially attached to the body and separates after several time steps. As the flow at the jet on the left side separates from the body, the flow around the apex also moves off the body (see Figure 11). This time delay before ventilation is not predicted by the present theory. The theory assumes the flow is similarity flow when determining the initial conditions. Similarity flow requires motion that is time independent. In particular, the determination of the wetted points is based on the fluid already being in motion. In reality the water needs to be accelerated from rest. The initial impact is not similarity type flow until the water has accelerated. The video pictures for Type B impact show the flow stays attached for only the first few frames. The distance covered by the body is on the order of 0.1 inch (2.5 mm) and the time is approximately 0.001 second. As the Type B impact continues the flow exhibits highly non-similar behavior near the apex. In some cases the free surface forms a jet that initially moves along the body and then detaches, but remains jet-like (see Figure 11). In other cases the free surface does not form a jet and remains flat as the body moves away (see Figure 12). In Figure 12 the flow near the apex on the left side curves out from the body. The connection between the flow curving away from the apex and the free surface near the jet often becomes indistinct (the reason for this is not clear).

$$\theta = 15^{o}$$

$$\theta = 20^{o}$$

$$\theta = 25^{o}$$

$$\theta = 30^{o}$$

$$\theta = 34^{o}$$

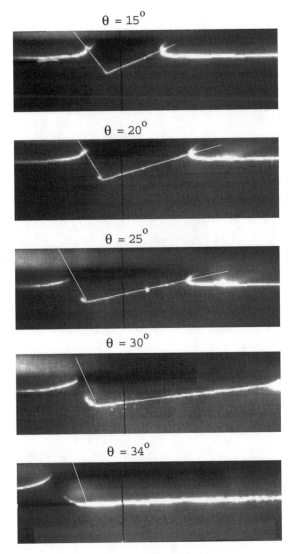

Figure 10. Transition from Type A to Type B Impact with different θ's and U/W = 0.75.

Of special interest is the case of small asymmetry and large horizontal impact velocity. In this situation the ventilation is mainly due to the horizontal velocity component. Figure 12 shows an example of this kind of ventilation. Initially the flow is attached on both sides of the body, but quickly the flow near the apex on the left side moves from the body. The free surface on the left side does not actually form a jet since the ventilation

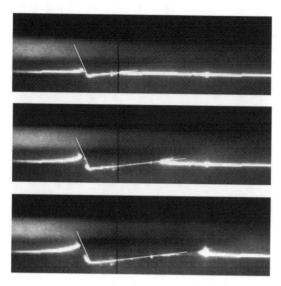

Figure 11. Type B Impact with $\theta = 30°$ and U/W = 0.75.

is caused by the body moving away too quickly. A jet on this side cannot form until the flow around the apex reaches the surface. In this situation, the behavior of the free surface on the left side is very different than for a Type A impact. What exactly is happening is difficult to determine from the video records and requires further study to be understood properly.

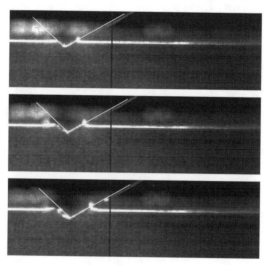

Figure 12. Type B Impact with $\theta = 5°$ and U/W = 2.0.

In summary, this experimental investigation of initial flow separation from the apex revealed many interesting flow characteristics. Most of these characteristics are not entirely understood or explained by any existing theory. Further study of this flow and the development of a flow model is an important step in understanding impact with asymmetry and horizontal impact velocity.

CONCLUDING REMARKS AND FUTURE DIRECTION

The two-dimensional impact theory presented here is motivated by the need for analytic design tools for predicting high speed planing dynamic stability. The two-dimensional asymmetric impact problem with horizontal impact velocity is of particular importance in studying horizontal plane dynamic stability. The model described here captures the relevant details of the impact flow as well as jet-formation. Initial separation of the flow from the keel due to asymmetric impact or horizontal impact velocity is examined in relation to the present theory. The impact model described can give the hydrodynamic characteristics, including the transient impact loads, of a vessel section impacting with asymmetry and horizontal impact velocity.

Experimental drop tests of a prismatic wedge were performed to gain understanding of initial water impact when asymmetry and horizontal impact velocity are involved. In particular, the initiation of Type B flow was investigated. The predictions for $c_2{}^*$ were compared between the theory and experimental results. The ratios of the jet spray-root locations, $c_2{}^*$, predicted by the method presented in this paper were compared with the experimental data. Good agreement between the data and the numerical predictions was shown for small degrees of asymmetry and small ratios of horizontal to vertical impact velocity. The numerical prediction of the transition point from Type A to Type B flow compared very well with the experimental observations. The flow behavior just before and at separation is very complex and is not incorporated in this impact model. This is clearly an area requiring additional research.

The results of an impact model that included both asymmetry and horizontal velocity could be incorporated into a nonlinear motion simulator in order to provide an analytical transverse stability tool. The results would allow for a dynamic righting arm curve to be developed for high speed planing craft. Horizontal velocity is a significant component of transverse planing stability and therefore needs to be addressed. That is the goal of developing the model presented here.

ACKNOWLEDGMENTS

This research has been supported by the Department of Defense, Office of Naval Research and the Link Foundation. The drop tests described here were done in the spring of 2000 in the Marine Hydrodynamics Laboratory at the University of Michigan, Ann Arbor with the help of Luke DeVito, Kevin Maki and Joe Krasny.

REFERENCES

[1] T. von Karman. "The impact of seaplane floats during landing." Technical Note 321, NACA, Washington, D.C. (1929).
[2] H. Wagner. "Uber stross-und gleitvorgange an der oberflache von flussigkeiten." ZAMM, 12:193–215 (1932).
[3] W. S. Vorus. "A flat cylinder theory for vessel impact and steady planning resistance." *Journal of Ship Research*, 40(2):89–106 (1996).
[4] L. Xu, A. W. Troesch, and W. S. Vorus. "Asymmetric vessel impact and planning hydrodynamics." *Journal of Ship Research*, 42(3):187–198 (1998).

Visualization of the Near-Boundary Hydrodynamics About Fish-Like Swimming Bodies

Alexandra H. Techet

Department of Ocean Engineering
Massachusetts Institute of Technology
Cambridge, MA 02139
Research Advisor: Dr. Michael Triantafyllou

ABSTRACT

Through particle image velocimetry (PIV) and Laser Doppler Velocimetry (LDV), the near boundary flow about fish-like swimming bodies is visualized and quantified in order to better understand the mechanisms behind the efficient propulsion of fish. Both a robotic swimming fish and a two-dimensional moving mat are investigated within. The robotic fish (MIT *RoboTuna*) is a complex three-dimensional flow problem. The waving plate, a reinforced Neoprene mat forced to oscillate by a piston driven system, is designed as a two-dimensional representation of the robotic fish. Swimming motion is modeled as a traveling wave with wavelength, λ, and varying amplitude, $A(x)$. Swimming parameters, such as Strouhal number and phase speed of the body wave, are varied to achieve different swimming configurations. Digital Particle Image Velocimeter (DPIV) of the robotic fish at Reynolds number 800,000 reveals a laminarization effect in the near boundary flow. Quantitative flow measurements near the waving wall illustrate a reduction in turbulence intensity with increasing phase speed.

INTRODUCTION

Engineers are forever on the quest of a faster, more efficient means of transportation, especially where underwater vehicles are used. The ability of a fish to propel itself over great distances quickly and gracefully, without apparent fatigue, is fascinating to engineers and biologists alike. Over millions of years, aquatic creatures have evolved to optimize their propulsive mechanisms in order to thrive in their specific environments. Fish, especially, seem to have perfected swimming, maneuvering and accelerating; yet the physics behind their seemingly effortless, efficient motion is not quite clear leading researchers to study the swimming and maneuvering of fish.

In order to further understand these amazing creatures, scientists have moved to study the muscles [1,2] and fluid dynamics of live swimming fish [3,4]. Scientists have captured these movements in robotic machines such as the MIT *RoboTuna* developed by [5]. [6] reviews current research into fish swimming.

To better understand the impressive swimming ability of these live and man-made creatures, the fish-fluid interactions need to be investigated further. Previously researchers have shown the ability of the caudal fin of a fish to produce a jet-like wake similar to that of a flapping foil [7]. However this phenomenon does not completely explain the high propulsive efficiencies of the swimming fish.

Experimental results revealing the swimming hydrodynamics of live fish can be difficult to obtain in laboratory environments. As a result, Barrett constructed a robotic underwater flexible hull vehicle in the shape of a tuna with a lunate tail capable of swimming in a straight line on a towing carriage, and on which drag force and energy analysis could be performed [5]. It was shown experimentally that over narrow ranges of various swimming parameters, the robot could achieve extremely high efficiencies, seemingly higher than those of conventional marine propulsors, and in addition, can experience lower drag than a rigid-body hull [8,9].

The advent of more flexible flow visualization techniques has given way to better investigations into the swimming of live fish. Anderson [4] studied the Giant Danio in a still water tank using Digital Particle Image Velocimeter (DPIV) and high-speed camera imaging in Kalliroscopic fluid. Gray [10] compared these experimental results with numerical simulations of the same fish. These studies yielded valuable information into the wake features of the fish but due to the processing techniques used, offered little information on the near boundary features of the swimming fish.

Like the fish body motion, a wave traveling down a flat plate produces an undulating motion that will affect the fluid near the plate boundary and also the vortex shedding mechanisms. The evolution of turbulence is of special interest as the undulatory boundary motion has been found to potentially result in suppression of turbulent structures and reduction of wake signatures [11,12]. The swimming plate offers a controlled two-dimensional experimental setting to further study the effects of an undulating motion on a boundary layer and the opportunity to draw a parallel with the hydrodynamics of swimming fish.

Taneda and Tomonari investigated the flow around a flexible, waving mat, revealing a unique effect as the wave phase speed increased beyond the free stream velocity [13]. It was observed that the flow began to re-laminarize at the crests, and then eventually over the entire mat (troughs and crests), as the phase speed approached and then exceeded the velocity of the free stream. This result has an interesting application to the swimming fish, since the motion of its body is essentially that of a traveling wave.

It has been shown, in other cases as well, that active flow control can be realized through unsteady motion [14,15]. [16] and [12] have shown that oscillatory rotation of a circular cylinder can effectively control separation and reduce the wake signature and thus the drag on a body. The unsteady undulating motion of the waving plate tended to reduce the drag wake signature as well as reduce near body turbulence [11,13]. These studies of unsteady motion for the purpose of flow control give valuable insight into the area of unsteady flow control. The problem of the waving plate, or swimming fish, is one more extension of these ideas.

The objective of this work is to conduct an experimental study of the hydrodynamic characteristics of unsteady flow near the fish-like swimming boundaries and the separation (or lack thereof) of the fluid. How the undulating motion of the boundary affects both the boundary layer development and vortex shedding off the trailing edge is of great interest, but only the former will be addressed herein. A correlation between this problem and that of a swimming fish will also be discussed, using visualization results obtained from a study of the MIT *RoboTuna*.

EXPERIMENTAL SETUP

The evolution of videographic equipment, digital computers and high power pulsed lasers has brought Particle Image Velocimetry, PIV to the forefront of fluid mechanic experimentation and visualization. A comprehensive review of PIV and particle imaging techniques is given by Adrian [17-19]. The digital implementation of PIV is outlined by Willert and Gharib [20].

PIV is the measurement of the instantaneous whole-field velocity components and the respective vorticity field, by comparing snapshots of particle-laden flow at two instances close in time. The movement of the particles from image to image yields the velocity vectors associated with the flow. In general PIV track groups, patterns, of particles in an image using digital FFT and cross-correlation routines, while particle tracking velocimetry (PTV) tracks each individual particle.

The near-body flow visualization on the MIT *RoboTuna* was performed in the MIT Testing (Towing) Tank. The tank is 100 ft long, eight feet wide and four feet deep on average. The MIT *RoboTuna* is a biologically inspired robot animated through a series of tendons and pulleys. The RoboTuna is modeled after the Bluefin Tuna with a body length of 1.25 meters and tail fin span 0.3 meters. The robot's swimming motion is dictated by a traveling wave along backbone. The motion of the backbone, *y(x,t)*, is given by

$$y(x,t) = a(x) \, sin(kx-\omega t) , \tag{1}$$

where the amplitude *a(x)* is

$$a(x) = c_1 x + c_2 x^2 , \tag{2}$$

where k is $2\pi/\lambda$ and ω is the frequency; c_1 and c_2 are constants. The wave phase speed is defined as $Cp = w/k$. The robot's swimming motion was optimized by genetic algorithm in order to obtain optimal swimming configurations. A lengthy discussion of the robot's operating parameters is found in Barrett's theses [5] and [8].

DPIV studies of the *RoboTuna* were performed with a 400 W/pulse Spectra Physics dual pulsed ND:Yag laser specifically designed for PIV. The laser light sheet was formed and passed through a clear window in the side of the tank. The light sheet was wider than necessary to allow for longer image sequences as the CCD camera passed over the sheet. The camera was positioned at one point on the robot, aft of the streamlined mast, looking down into the water, partly submerged in a watertight housing to avoid surface distortion to affect the images. A schematic of the experimental setup can be seen in Figure 1.

The black and white CCD video camera used was a Texas Instruments MULTICAM MC1134P. It records the flow at a maximum pixel resolution of 752 x 480 pixels and standard frame rate of 30 *Hz*. A zoom lens is employed to obtain a small field of view, 2.0 *cm* x 3.0 *cm*, near the body boundary. An EPIX frame grabber digitizes the images in real-time into the memory of a Pentium 133 MHz PC.

The *RoboTuna* images were sparsely seeded due to difficulty in seeding the large tank volume sufficiently. In order to perform DPIV processing it is necessary to have a more densely seeded flow field, thus Particle Tracking was employed. PTV code, specially designed at MIT to take into account the presence of a boundary image, was used to analyze the sequences. Limited phase match averaging was performed on the data sets.

In order to study the hydrodynamics of a waving plate, an apparatus

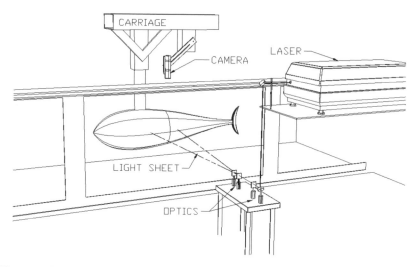

Figure 1. Experimental setup for the DPIV experiments on the MIT *RoboTuna*. The dual-pulsed ND:Yag laser is mounted on the wall to minimize vibrations. The laser beam is steered through the optics and enters the tank in a horizontal sheet through a clear window. The beam is located at the lateral (mid-) line of the fish. The camera is attached to the carriage, is partially submerged in a watertight housing, and looks down on the horizontal laser sheet.

was designed and built to operate in the hydrodynamics tunnel at MIT. The apparatus is an eight link piston-driven system similar to a straight-eight car engine. Solid modeling of the waving plate apparatus was done with SDRC I-DEAS before any machining took place. An isometric three-dimensional view of the mechanism inside the test section of the hydrody-namics tunnel is shown in Figure 2. This model illustrates the crank-piston mechanism that drives the waving mat.

Using the MIT *RoboTuna* [5] and [11] as a guide, two separate backbone shapes were chosen for the waving plate motion. The first shape had am-plitudes on the order of the MIT RoboTuna. However, since the mat is po-sitioned in the water tunnel slightly closer to the top wall than the bottom, such amplitudes caused flow problems and mat flutter due to the constric-tion of the flow when the trailing edge was at its maximum amplitude of +4.0 inches. The amplitude of this motion was dictated by $a(x) = 0.1\,x$, where x is distance from the leading edge of the mat, thus piston 1 is at $x=5$ inches and piston 8 at $x=40$ inches. It would be wise to mention here that the lead-ing edge of the neoprene was 5 inches aft of the actual leading edge of the support section. The second mat shape was more similar to [11] with an amplitude $a(x) = 0.75/12\,x$. This smaller amplitude alleviated the flow con-striction near the top of the tunnel and resulted in a smooth mat motion.

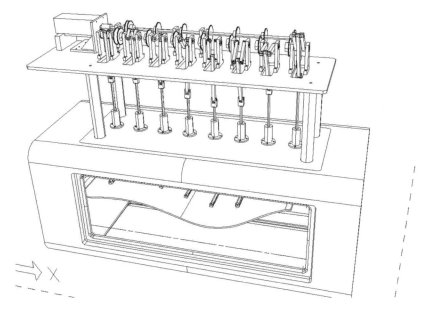

Figure 2. Solid modeling of the waving plate apparatus showing the drive mechanism and the piston configuration. A phase-offset of the crank arm motion, from the previous piston, dictates the shape of the traveling wave in conjunction with the length of each set of crank arms. Chain links and sprockets allow a common drive shaft to power each individual crank-piston mechanism with one single DC motor.

A linkage arm connects each 3/8–inch diameter piston rod to the crank arms. Each set of crank arms is individually driven by a 0.25–pitch sprocket connected to the drive sprockets with 0.25–pitch metal chain, similar in style to standard bicycle chain. The 1/2"-diameter drive shaft is attached through a flexible coupling to a 1/3–hp DC motor. The Bodine Model 42A7BEPM-5H DC motor is capable of maximum torque loads up to 87 lb-in and speeds up to 139 rpm.

The pistons are 5" apart and guided through the tunnel window by two Frelon lined linear bearings. There is no seal around the piston. The only leakage experienced was the small amount of water that seeped out with the motion of the pistons through the bearings. This water amounted to a few teaspoons of water per day and thus was not a concern. There is a space under the bearing mount that would allow for the insertion of an o-ring if necessary. Ultimately this o-ring was omitted in order to reduce the friction on each piston.

From the top of the pistons, a linkage arm connects to the crank mechanism that controls the mat motion. The link arms are fitted with a needle roller bearing at each end, which rolls directly on a 3/8″ shaft. The link arms are connected at the top to two crank arms. Both crank arms rotate in ABEC-3 bearings mounted in 4″-high shaft supports. One of the arms is connected to a 0.25–pitch (2″ diameter) sprocket that is linked by chain to the drive shaft sprockets. The drive sprockets are the same diameter as the crank sprockets. Different sprockets could be attached to change the available torque/speed range of the system, if necessary.

The actual mat is made from 1/4″ high-grade Neoprene sheeting. The rubber mat is reinforced with square, hollow brass rods glued to the surface with compliant rubber cement. The bond is made more secure with a fillet of clear silicone RTV and a bolt at each end to prevent the rods tearing off. The bolts at the end are 100° flat head screws that are flush with the surface on the flow-side of the mat. The bottom end of each piston hinges around a cross-rod that is encased, at either end, in sliders to allow the mat to flex without stretching or compressing as the pistons move up and down. Small stainless steel springs help the cross-rods return to a neutral position and allow the mat to move smoothly.

The leading edge of the rubber mat is attached to an aluminum support. This support is made from a half-inch thick, horizontal aluminum plate, with a rounded, smooth leading-edge, that spans the width of the test section. This bottom plate is welded to two NACA 0012 shaped, vertical foil sections and a smaller horizontal plate on top of the foil sections that has two tapped hole that allow the leading edge support to be firmly bolted into the test section. This allows for a five-inch, flat, smooth plate, with a rounded leading edge, to split the incoming flow just before the mat and zero motion at the leading edge of the rubber.

An angle offset between each of the crank arms creates the traveling wave effect. This angular offset resulted in a wavelength, λ, 0.8 times the length of the mat, L $(L = 1.25\lambda)$. Since the total length of the mat, including the leading edge plate and the trailing edge extension beyond the last piston, is 49″ (or 1.25 meters), $\lambda = 1.0$ meters. The motion of the mat, using the equation for amplitude defined above, is $y(x) = A(x) \sin(kx—wt)$, where k is the wave number, $k = 2\pi/\lambda$, w is the piston frequency in radians, and t is time in seconds. The phase speed, Cp, of the mat can be deduced from simple wave theory and written as

$$Cp = \omega/k = (2 \pi f)/(2 \pi / \lambda) = f \lambda. \tag{3}$$

Since the piston motion dictates the frequency of the wave, the phase speed is intrinsically linked to the rotational speed of the motor. The addition of the linear motion potentiometers (LMP's) alleviated the need for a motor encoder to deduce the motion frequency although a simple encoder was added for the LDV experiments. The frequency of the piston motion was recorded by the DasyLAB data acquisition software on a DAS16 Jr. card, from Computer Boards, wired to measure 16 channels single-ended. The LMP signals were processed in MATLAB prior to each run to ensure repeatability in the motion. The frequency of motion was correlated with the motor-controller potentiometer output voltage to ease setup. Frequencies were chosen based on the ratio of phase speed to free-stream flow Uo. Five such ratios were chosen:

$$\zeta = Cp \,/\, Uo = 0.3, 0.6, 0.8, 1.0, 1.2. \qquad (4)$$

Using equation 3 the requisite frequency is thus $f = \zeta\, Uo \,/\, \lambda$.

Tests on the waving plate apparatus are performed in the MIT Hydrodynamics Lab. The lab has a 33,000 gallon re-circulating water tunnel that was designed originally to test propellers. It has a square test section that is 20" x 20" in cross section and 45" long. The tunnel is capable of sustaining flow speeds up to 20 ft/s. A photograph of the apparatus atop the test section is shown in Figure 3.

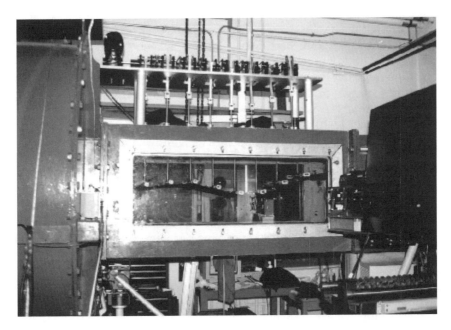

Figure 3. Waving plate apparatus assembled in the MIT Hydrodynamics tunnel. The drive mechanism sits atop the test section. Through the clear side window the mat can be seen. The digital CCD camera used for DPIV is placed on the X, Y, Z-traverse (at right) that also carries the LDV system.

On all four sides of the test section is a removable window that allows access to the inside of the tunnel. The top opening in the test section is fitted with the base of the waving plate apparatus. The bottom window is a clear, laser-quality polycarbonate window. The front window (on the traverse side) is another laser-quality window, and the rear window is covered on the outside by black cloth to prevent the light from reflecting off the wall of the room and also for safety. This window is also fitted with a 3/8" shaft that has a ruler attached to the end inside the tank used to calibrate the PIV images. The ruler stays inside of the test section, flush to the walls during testing. When a calibration image is taken, the ruler is moved into the plane of the laser sheet and digitized in both a horizontal and vertical direction.

The laser used in the DPIV experiments on the waving plate was located below the test section on a breadboard with optics to create the light sheet. First the beams exit the head to the New Wave Gemini:PIV laser and travel through a convex lens to collimate the beam. Next the beam hits a 45° first surface mirror that steers the horizontal beam vertically. The beam now passes through a convex lens that focuses it to a line and then through another convex optic, positioned 90° to the second optic, which spreads the thin line to a thin sheet. The light sheet is aligned in the direction of the flow and illuminates the mid-plane of the waving mat.

Timing of the system was done through the National Instruments timing board and software from General Pixels. Due to technical difficulties, an external timing box from Berkeley Nucleonics was used in addition to the General Pixels timing box. The timing card and software, in synch with the frame marker of the Pulnix camera, output the external trigger to the Berkeley timing box which in turn creates the pulses.

The Pulnix TM-1040 1K x 1K CCD digital camera is mounted on a x,y,z traverse (seen at the right in figure 3) which allows for accurate positioning, and re-positioning, of the field of view. The traverse accuracy is ±0.1 mm. Video are stored in 356 Mb of computer memory and transferred to hard drive. Image sequences are analyzed by DPIV software developed by S. McKenna of Woods Hole Oceanographic Institution.

DPIV of the waving plate experiments at a 1Kx1K resolution presented an interesting data storage and processing problem. There was close to 100 GB of data acquired, something that would have been very expensive only a few years ago and was still daunting today. Processing the images was also a time consuming process. It would be to our advantage to be able to synchronize the data acquisition with the mat motion, but we save this for the next round of experiments.

LDV data also was taken on the waving mat in the hydrodynamics tunnel. The LDV system consisted of a 6–Watt Argon Ion laser and optics by TSI, Inc. The data acquisition system from Dantec allowed for u-and v-velocity data to be recorded and ordered in angle bins averaged by phase over 360 degrees of mat motion for one fixed point. The x-y-z traverse allowed a cut of data to be taken at any point along the mat length.

For the purpose of these experiments, the laser was positioned 30 *cm* downstream of the leading edge. Data was taken along a vertical cut, from the top extreme of the mat to one cm below the mat's lowest point of travel, every 2 *cm* with U-component only and then again every 4 *cm* with U and V channels both turned on. Since the V-component is measured at the intersection of two beams aligned in the vertical plane, there is significant clipping of the upper beam near the mat. The beam crossing was located 1/3 of the distance in from the wall, out side of the tunnel wall boundary layer, in order to minimize the vertical distance of the measurement point from the waving mat.

An encoder, attached to the drive shaft, relayed the phase information to the Dantec acquisition system. The Dantec software ordered the raw data in angle bin from 0 to 360 degrees and phase averaged over each bin. It was specified that 36,000+ data points were to be acquired to assure approximately 100 samples per bin. Data rates varied depending on the flow speed—almost doubling for the 1 m/s case over the 0.5 m/s runs. The lowest acquisition rate was approximately 20 Hz when both U and V channels were turned on up to the highest at 180 Hz for U only. With both U and V channels active, the LDV system allowed for data to be acquired only at a time when both U and V components were deemed to be valid thus the significantly lower data rates.

NEAR BOUNDARY FLOW ABOUT MIT *ROBOTUNA*

Two swimming cases are targeted specifically to coincide with Barrett's power measurement findings of best drag reduction and best "self-propelled" swimming [5]. Through DPIV and particle tracking the flow in the boundary layer, down into the viscous sub-layer, can be resolved. Preliminary DPIV results confirm that laminarization of the boundary layer was realized for the case of the swimming *RoboTuna* [21,22]. Further experiments are scheduled to obtain more data in order to calculated ensemble averages and turbulence statistics.

Several runs were performed for straight drag (non-swimming) configuration at several points aft of the towing strut. These tests were used as a basis for comparison with the swimming cases. The straight-drag cases revealed a turbulent boundary layer profile over most the length of the fish until the near tail region where separation began. These tests were in stark contrast to the swimming cases studied.

For the swimming cases, two specific sets of parameters were chosen to correspond with Barrett's "Best Drag Reduction" and "Best Self Propelled"

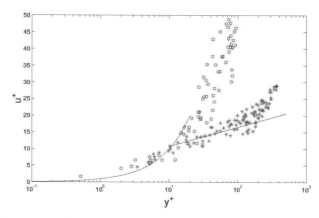

Figure 4. Comparison of a swimming and non-swimming case for the MIT *RoboTuna* "Best Self Propelled" swimming configuration. The asterisks indicate the non-swimming case and the stars the swimming case. The non-swimming case resembles a turbulent flat plate profile whereas the swimming case follows the laminar profile. The two solid lines are drawn as $u^+ = y^+$ and $u^+ = 2.5 \ln y^+ + 0.5$ representing a turbulent boundary layer on a flat plate [23]. The experimental data is non-dimensionalized by the frictional wall velocity, which is calculated as the slope of the velocity data at the boundary. Data is taken just aft of the anal fin and at the mid-line of the robot.

cases. The result presented here are for the BSP case where $U = 0.7$ m/s (Re $= 875,000$), St $= 0.182$, and Cp/Uo $= 1.14$. For this case it can be seen in Figure 4 that the swimming configuration tends to a laminar profile whereas the non-swimming case resembles a turbulent boundary layer on a flat plate. The data was converted to wall coordinates using the slope of the velocity near the wall to determine a value for the frictional velocity. While the waving plate is not the same as a flat plate, flat-plate boundary layer theory is used as a guide to better understand the data obtained.

LDV AND DPIV OF THE FLOW NEAR A WAVING MAT

Tests on the waving plate are performed in the MIT Propeller Tunnel. The tunnel test section has dimensions of 20 in. x 20 in. across and is 44 in. long. Flow speed is variable up to 20 ft/s.

Flow visualization of the boundary layer around the swimming plate is performed at various Reynolds numbers up to 1,000,000. Presented here are for Reynolds numbers 500,000 and 1,000,000. Both LDV and DPIV tests were performed.

The LDV data was obtained and phase averaged over a cycle by input-

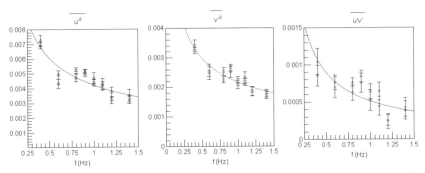

Figure 5. LDV results from the waving plate under piston #6 at the crest of the motion at $Re_\lambda = 10^6$ ($U_o = 1.0$ m/s). This plot shows the reduction in turbulence intensity as the phase speed of the traveling wave increases ($C_p = f\lambda$; $\lambda = 1.0$ m). The data is averaged over 100–150 cycles of motion and two sets of data are presented for each phase speed to indicate repeatability. The error bars represent the random error of the phase averaging.

ting an encoder signal to the Dantec processor. This inductance pickup was synched with the point of interest along the mat. Data presented here was acquired under piston #5, or 25 inches down the mat. It was found that at the trough point of the mat motion, the turbulence intensity, $\overline{u'v'}/\overline{q^2}$, where $\overline{q^2} = \overline{U^2} + \overline{V^2}$, is reduced significantly. This is shown in Figure 5.

DPIV on the waving plate was done in two different configurations, first with an eight centimeters squared field of view (FOV) and later with a two centimeter FOV. The first configuration yielded lower spatial resolution due to the large FOV. By reducing the FOV we were both able to gain higher spatial resolution and also obtain more data very close to the mat. The seeding density was sufficient to process the smaller FOV images with a 16x16 box, versus the 64x64 box for the larger FOV. This also dramatically improved our spatial resolution.

DPIV images will be processed for comparison with the LDV Results. Preliminary analysis indicates the results agree with those of the *RoboTuna* study.

CONCLUSIONS

In keeping with the results by [11] and [13] the LDV and PIV data on the waving plate and robotic fish indicate a reduction in turbulence near the body for fish-like swimming motions. These results agree well with the numerical simulations of a two-dimensional waving mat in [24]. The dramatic reduction in turbulence intensity at the trough and crest of a waving mat was found through the simulations for increased wave speeds around $Cp/U = 1.12$. It was also shown that the boundary layer profile changed significantly with phase speed. The case of the waving plate offers us a good vehicle to understanding the flow about swimming fish, but warrants further investigations in the future into the effects of different amplitude to wavelength ratios as well as the effects of frequency. There are many different fish and swimming creatures that could hold new and exciting information for biomimetic propulsion mechanisms.

ACKNOWLEDGEMENTS

I would like to acknowledge the generous support of the Link Foundation Fellowship in Ocean Engineering and Instrumentation, in addition to the research support from the Office of Naval Research under grant N00014–00–1–0198. I would also like to acknowledge the use of the MIT Hydrodynamics Facility operated by Professor Jake Kerwin and the help of Mr. Richard Kimball in the tunnel operation.

REFERENCES

[1] J. Gray. Studies in animal locomotion: VI. The propulsive powers of the dolphin. *Journal of Experimental Biology*, 13(2):192–199 (1936).

[2] J. Gray. *Animal locomotion*. Weidenfield and Nicolson, London (1968).

[3] E. J. Anderson, A. H. Techet, W. R. McGillis, M. A. Grosenbaugh, and M. S. Triantafyllou, "Visualization and analysis of boundary layer flow in live and robotic fish" *First International Symposium on Turbulence and Shear Flow Phenomena*, Santa Barbara, CA, USA, 12–15 (1999).

[4] J. Anderson. *Vorticity control for efficient propulsion*. PhD dissertation, Massachusetts Institute of Technology and the Woods Hole Oceanographic Institution, Department of Ocean Engineering (1996).

[5] D. Barrett. *The design of a flexible hull undersea vehicle propelled by an oscillating foil.* Master's thesis, Massachusetts Institute of Technology, Department of Ocean Engineering (1994).

[6] G.S. Triantafyllou, M.S. Triantafyllou, and D.K.P. Yue. Hydrodynamics of Fishlike Swimming *Annual Rev. Fluid Mech.*, Vol. 32: 33–53 (2000).

[7] M. Triantafyllou and G. Triantafyllou. An efficient swimming machine. *Scientific American*, 272(3):64–70 (1995).

[8] D. Barrett. *Propulsive efficiency of a flexible hull underwater vehicle*. PhD dissertation, Massachusetts Institute of Technology, Department of Ocean Engineering (1996).

[9] G. Triantafyllou, M. Triantafyllou, and M. Grosenbaugh. Optimal thrust development in oscillating foils with application to fish propulsion. *Journal of Fluids and Structures*, 7:205–224 (1993).

[10] M. Wolfgang, J. Anderson, M. Grosenbaugh, D. Yue, and M. Triantafyllou. Near-body flow dynamics in swimming fish. *Journal of Experimental Biology* (1999).

[11] J. M. Kendall. The turbulent boundary layer over a wall with progressive surface waves, *J. Fluid Mech.*, 41:259–81 (1970).

[12] P. Tokumaru and P. Dimotakis. Rotary oscillation control of a cylinder wake. *Journal of Fluid Mechanics*, 224:77–90 (1991).

[13] S. Taneda and Y. Tomonari. An experiment on the flow around a waving plate. *Journal of the Physical Society of Japan*, 36(6):1683–1689 (1974).

[14] J. Ffowcs-Williams and B. Zhao. The active control of vortex shedding. *Journal of Fluids and Structures*, 3:115–122 (1989).

[15] R. Gopalkrishnan, M. Triantafyllou, G. Triantafyllou, and D. Barrett. Active vorticity control in a shear flow using a flapping foil. *Journal of Fluid Mechanics*, 274:1–21 (1994).

[16] S. Taneda. Visual observations of the flow past a circular cylinder performing a rotary oscillation. *J. Phys. Soc. Japan.* 45:1038–43 91 (1978).

[17] R. Adrian. Multi-point optical measurements of simultaneous vectors in unsteady flow: A review. *International Journal of Heat Fluid Flow*, 7:127–145 (1986).

[18] R. Adrian. Particle-imaging techniques for experimental fluid mechanics. *Annual Review of Fluid Mechanics*, 23:262–304 (1991).

[19] R. J. Adrian, R. D. Keane, and Y. Zhang. Super-resolution particle imaging velocimetry. *Meas. Sci. Tech.*, 6:754–768 (1995).

[20] C. E. Willert and M. Gharib. Digital particle image velocimetry. *Exps. Fluids*, 10:181–193 (1991).

[21] A. H. Techet, E. J. Anderson, W. R. McGillis, M. A. Grosenbaugh, and M. S. Triantafyllou, "Flow visualization of swimming robotic fish in the near boundary region" *Third International Workshop on Particle Image Velocimetry*, Santa Barbara, CA, USA, 16–18 September 1999).

[22] A.H. Techet and M. S. Triantafyllou, "Boundary layer relaminarization in swimming fish" *Proc. ISOPE 1999*, Brest, France, 30 May—4 June 1999).

[23] Tennekes and Lumley. *A First Course in Turbulence.* The MIT Press, Cambridge, MA (1972).

[24] Zhang X. *Direct numerical simulation of the flow over a flexible plate.* PhD thesis. MIT, Cambridge, Mass. (2000).